THE PALESTINE EXPLORATION FUND ANNUAL X

Series Editor

David M. Jacobson

Mapping Jordan Through Two Millennia

MAPPING JORDAN THROUGH TWO MILLENNIA

John R. Bartlett

PALESTINE · EXPLORATION · FUND · 1865 ·

Routledge
Taylor & Francis Group
LONDON AND NEW YORK

2008

First published 2008 by Maney Publishing

2 Park Square, Milton Park, Abingdon, Oxfordshire OX14 4RN
52 Vanderbilt Avenue, New York, NY 10017

Routledge is an imprint of the Taylor & Francis Group, an informa business

First issued in paperback 2019

ISBN 13: 978-1-905981-40-3 (hbk)
ISBN 13: 978-0-367-88590-8 (pbk)
ISSN 1753-9234

Cover image: Section of E. Reuwich's map showing the Sinai peninsula and Red Sea (from B. von Breitenbach's *Sanctarum Peregrinationum ... opusculum* [Mainz, 1486])

CONTENTS

LIST OF MAPS

COLOUR PLATES

(following p. 84)

FIGURES

PREFACE

This short study explores the history of the making of maps of the region now known as the Hashemite Kingdom of Jordan. To the biblical writers this region was known as 'the other side of the Jordan', and to mid-twentieth-century statesmen 'Transjordan'. These designations revealed that the region was for many defined by its position east of the River Jordan, and so east of the land of Israel and the 'Holy Land'. For Jewish and Christian readers of the scriptures, the lands east of the Jordan were on the whole enemy territory, the home of Edomites, Moabites and Ammonites, Midianites, Arabs and other unfriendly peoples. For Europeans of the Middle Ages and beyond, this region was known chiefly as the home of the Muslim. For European travellers of the sixteenth to early twentieth centuries, this region was under Ottoman rule and was not over-welcoming to foreign visitors. Through all these centuries the land west of the Jordan was much better known than the land east of it, and this fact is reflected in the travel accounts and maps that have come down to us.

The focus in this study is on the geographical area east of the River Jordan and the Wadi Arabah, approximately from Damascus in the north to the Gulf of 'Aqabah in the south. As is already clear, the changing political history and modern political sensitivities cause difficulties for any consistent naming of this area in a book such as this. To repeat the phrase 'the modern land of Jordan' every time it is needed is unnecessarily clumsy, and for brevity and reasonable clarity I shall use for the region the Latinised name 'Transjordan', the partner of the academic 'Cisjordan', with no intended reference (except where indicated) to the mandated territory created after the 1914–18 war.

The last two centuries have seen dramatic developments in the technology of mapmaking, to the benefit of modern maps of the land on both sides of the Jordan. There is now also an accompanying interest in the history of cartography, served by learned journals such as *Imago Mundi*, first published in 1935 under the editorship of Leo Bagrow. There has, unsurprisingly, been particular interest in the cartographical history of the Holy Land. R. R. Röhricht published in 1890 his *Bibliotheca Palaestinae*, with a section devoted to *Cartographia*. Map libraries have presented exhibitions of maps of the Holy Land, complete with published catalogues. In 1986 Eran Laor published an important volume, *Maps of the Holy Land: Cartobibliography of printed maps, 1475-1900* (New York: Liss; Amsterdam: Meridian). In the same year K. Nebenzahl published his own superbly illustrated and very useful *Maps of the Holy Land* (New York: Abbeville), and in 2001 Ariel Tishby edited *Holy Land in Maps* (Jerusalem: The Israel Museum). As can be seen from the titles, the focus and interest have been on the Holy Land, and not on the land east of the Jordan, except where its history can be seen through the eyes of the occupants of the west bank. In many studies the interest has tended to be on the details of publication history rather than on the actual content of the maps. The present study, however, is concerned not with the mapping of the Holy Land west of the Jordan, nor with the bibliographical details of publication-history, but with the history of the mapping of the land east of the Jordan and the development of the contents of the maps. A major interest is the identification on the modern map of sites referred to in ancient literature (this, after all, inspired much of the mapping of the last two millennia), and this study will therefore end with some note of the increasing amount of scholarly work in the twentieth century on the identification of biblical, classical and medieval sites. The detailed history of the maps themselves will end with the completion of the triangulated surveys of Transjordan in the 1930s; the study of more recent satellite mapping is for the next generation of cartographic scholarship.

A close study of the available maps of the land east of the Jordan from Roman times to the present day shows clearly which maps were based on first-hand experience of the land and which were based on biblical and classical scholarship. The Roman itineraries and the Peutinger Table had hard military reconnaissance and imperial administrative experience behind them; some medieval work shows information clearly drawn from the Crusaders' occupation of the land; but from the Renaissance until the early nineteenth century biblical scholarship rules supreme on the maps, much of what the Crusaders learned being forgotten until the appearance of Seetzen, Burckhardt, and Robinson and Smith and their mapmaker H. Kiepert. It was not until the work of the Palestine Exploration Fund and the army officers who worked under its aegis in the late nineteenth century that accurate maps bearing modern names in their correct location began to appear. It took the demands of war in the early twentieth century to complete the task of mapping the regions east of the Jordan.

Since the work of Nelson Glueck in the 1930s, and the establishment of the Hashemite Kingdom of the Jordan after the Second World War, there has been a steadily growing interest in the geography and history of the land east of the River Jordan for its own sake. In particular, there has been much archaeological activity led by the Department of Antiquities of Jordan. The history of the mapping of Jordan belongs to this field of enquiry as well as to the world of biblical studies; it is part of the history of the recognition of the place of this land in its wider context. Modern cartography pictures and records the geography of Jordan as precisely and carefully as it does that of any other part of the world's surface, its motives deriving from modern politics and economics rather than from devotion to the biblical narrative or the history of the Crusades. But the earlier motives were real and should not be despised; there were scholars in every age who found it important to attempt to draw maps, and it is important to give them due credit for their contribution, especially considering the limitations under which many of them worked.

I wish to thank a number of people for help in making accessible various books and maps, in particular Charles Benson and the staff of the Department of Early Printed Books in the Library of Trinity College, Dublin, Paul Ferguson and Simon Fernandez (Glucksmann Map Library, Trinity College, Dublin), Rupert Chapman and Felicity Cobbing (Palestine Exploration Fund, London), Haim Goren (Israel), John Sawyer (Perugia), Eileán Ni Chuillenáin (Trinity College, Dublin), Graham

Davies (School of Divinity, Cambridge), Joanna Snelling (Corpus Christi College Library, Oxford), and Jill Hughes (Taylorian Institute, Oxford). I thank Fr Michele Piccirillo of the Studium Biblicum Franciscanum, Jerusalem, the British Library, London, the Biblioteca Augusta Communale, Perugia, and the Board of Trinity College, Dublin, for permission to reproduce illustrations. I gratefully acknowledge my huge debt to Brendan Dempsey, Photographer in Audio-Visual and Media Service, Trinity College, Dublin, for his patience and skill in providing the photographic illustrations without which this work would be unintelligible. My thanks are due particularly to Liz Rosindale and Linda Fisher of Maney Publishing for their editorial assistance and expertise, without which this book would not exist, and to the Publications Sub-Committee of the Palestine Exploration Fund which accepted this work for publication. I owe especial thanks to my wife, Olivia, for her constant stimulus and encouragement through the last five years.

JOHN R. BARTLETT
November 2007

ABBREVIATIONS

ACOR	American Center of Oriental Research
ADAJ	Annual of the Department of Antiquities of Jordan
AJP	American Journal of Philology
ASOR	American Schools of Oriental Research
GJ	Geographical Journal
IEJ	Israel Exploration Journal
IM	Imago Mundi
JHS	Journal of Hellenic Studies
JRA	Journal of Roman Archaeology
JRGS	Journal of the Royal Geographical Society
JSOTsup	Journal for the Study of the Old Testament: Supplement Series
JTS	Journal of Theological Studies
PEF	Palestine Exploration Fund
PEFQS	Palestine Exploration Fund Quarterly Statement
PEQ	Palestine Exploration Quarterly
PPTS	Palestine Pilgrims' Text Society
QDAP	Quarterly of the Department of Antiquities in Palestine
SBF	Studium Biblicum Franciscanum
SHAJ	Studies in the History and Archaeology of Jordan
TAVO	Tübinger Atlas des Vorderen Orients
ZDPV	Zeitschrift des deutschen Palästina-Vereins

CHAPTER ONE

WESTERN KNOWLEDGE OF THE TRANSJORDAN REGION IN THE FIRST MILLENNIUM AD

Systematic exploration by European scholars of the regions east of the Rift Valley line marked by the upper Jordan river, the Sea of Galilee, the lower Jordan, the Dead Sea and the Wadi Arabah did not begin until the nineteenth century. The work of Seetzen and Burckhardt in the first two decades of the nineteenth century, the travels of Robinson and Smith in the middle of the century, and the surveying work of the Palestine Exploration Fund in the final decades led to the military surveys of the twentieth century and the development of modern mapping, using first aerial photography and then satellite imaging to provide maps of an accuracy unimaginable in earlier centuries. Before the nineteenth and twentieth centuries, however, mapmakers concentrated on 'the Holy Land', drawing their information mainly from the Bible and the classical writers, and until the eighteenth century at least presenting the regions beyond the Jordan mainly in so far as they related to the biblical story.

The first non-biblical information about the lands east of the River Jordan comes from Roman sources. The early third century *Itinerarium provinciarum Antonini Augusti* gives us some place names on a route from Heliopolis (Baalbek) south via Abila, Damascus, Aere, Neve, Capitoliada, Gadara and Scythopolis, but names nowhere south of Scythopolis. The *Itinerarium Burdigalense* (Bordeaux Pilgrim, a.d. 333) also mentions Scythopolis, but otherwise tells us nothing about Transjordan. (For texts of these itineraries, see Ortelius [1600], and Cuntz [1929].) Much more informative are the late fourth-century–early fifth-century *Notitia Dignitatum*, 'list of office-holders', and the perhaps fourth-century road-map of the Roman empire known as the *Tabula Peutingeriana*, or Peutinger Table.

Of the *Tabula Peutingeriana* (Pl. 1A), discovered *c.* 1500 by the German poet Konrad Celtes but named after the Augsburg scholar Peutinger to whom he entrusted it for publication, D. F. Graf (1995, 143) comments that 'It is generally recognised that this medieval manuscript is based on a prototype of an itinerary that dates from the early Roman imperial period'. G. W. Bowersock (1983, 185) believes that the ultimate source was the first-century BC map made by Marcus Agrippa (64/3–12 BC) for Augustus and displayed at Rome in the Porticus Vipsaniae. Whatever its origins, 'its main purpose was the reproduction of the land routes of the known world between Britannia and Taprobene (Sri Lanka)' (Zimmermann 2003, 1140). This document is therefore clearly important evidence for knowledge of the routes of the Roman Empire at an early stage of its development. For the Transjordan, this map shows a route starting from Tiberias on the west side of the Lake of Tiberias (Sea of Galilee), crossing eastwards to Bostra, and then running south via Philadelphia (Amman) to Haila on the Gulf of 'Aqabah. This road is apparently joined at Hatita (Kh. Samra': Desreumaux and Humber 1981, 35) by a road coming from Damascus in the north, and at or near Rababatora by a road coming from Thamara to the west. At Haila it is met by a road running south across the Negev from Oboda. This map shows the main towns and other stations on the roads, with the distances between them in Roman miles.

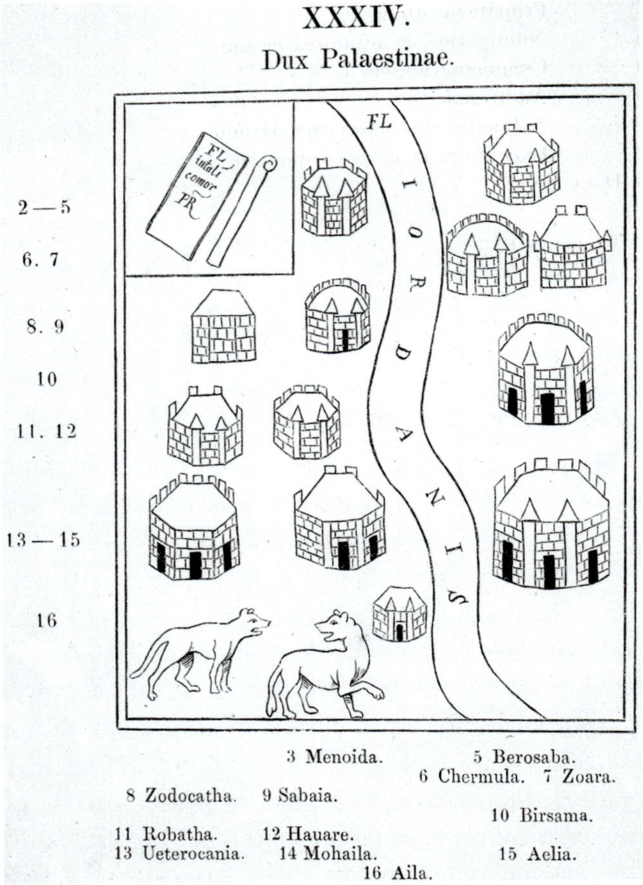

FIGURE 1. Illustration from fifteenth-century manuscript of *Notitia Dignitatum* (Seeck 1876). (*The Board of Trinity College, Dublin*)

The *Notitia Dignitatum* is 'a list, as its name implies, of civilian and military office holders and administration, accompanied by maps of appropriate locations', dated between 395–413, perhaps with later revision (O. A. W. Dilke 1985, 167–69). The extant manuscripts (held at Oxford, Paris, Munich, Cambridge and Frankfurt) are all comparatively late, from the fifteenth century, probably copied from a tenth-century Codex Spirensis, from Speyer Cathedral (Harley and Woodward 1987, 244–45). The maps 'seem to have been compiled line by line from the text, by someone who had little topographical knowledge and who inserted place or province names in textual order', and were possibly 'invented in the Dark Ages' (Dilke, ibid.; cf. Harley and Woodward 1987, 244). Many of the place-names west and east of the River Jordan mentioned by the *Notitia Dignitatum* (Fig. 1) appear also in the Peutinger Table, and between them these documents give some idea of Transjordan as it was known, at least to the Roman military and civil administration, in imperial times, the *Tabula Peutingeriana* showing the major route through Transjordan (marking the more important towns Bostra, Philadelphia and Petra with a conventional symbol of two towers) and the *Notitia Dignitatum* listing the main military and administrative centres. The *Notitia Dignitatum* gives a number of place-names which do not appear on the Peutinger map. The table below lists the places on the Peutinger map's routes through Transjordan, from north to south, alongside equivalents from the *Notitia Dignitatum*, and also from Ptolemy (see Ch. 3 below), from Eusebius' *Onomasticon* and from the Madeba Map.

The Peutinger map raises some interesting questions. The route from Tiberias travels roughly east via Gadara (Umm Qeis), Capitolias (Beit Rās) and Adraha (Dera'ā) to Bostra (Bosrā), where it turns sharply south-west towards Philadelphia (Amman) via Thantia (Umm el-Jimāl?) and Hatita (Kh. Samrā'). One would expect the road south from Damascus to Philadelphia to meet the road from Tiberias at Bostra, but in fact according to the Peutinger map the road from Damascus is routed via Aenos, Canatha (Qanawāt) and Rhose (between Bostra and Salecah [Salkhad], Bowersock 1983, map page 101). The Peutinger map road from Damascus to Philadelphia seems to have been routed east of Jebel ed-Drūz; the later Roman road across the Leja' west of Jebel ed-Drūz 'had not been built at the time of the Peutinger Table's archetype' (Bowersock 1983, 177).

The *Tabula Peutingeriana* and Eusebius' *Onomasticon* present different pictures of the Transjordan. The Peutinger map limits itself to the main routes running through Transjordan. Eusebius mentions the important places on them — Kanatha, Adraa, Bostra, Gadara, Philadelphia, Areopolis, Petra and Aila — but his interest is in the wider range of biblical and ecclesiastical sites, including such places

Tabula Peutingeriana	*Mileage to next station*	*Notitia Dignitatum*	*Ptolemy*	*Eusebius Onomasticon*	*Madeba Map*	*Modern site*
Tiberias	16					
Gadara	16		Gadara	Gadara		Umm Qeis
Capitoliada	16		Capitolias			Beit Ras
Adraha	24		Adra	Adraa		Dera'a
Bostris	24	Bostra 37.21	Bostra	Bostra		Bosra
Thantia	9	Thainatha 37.29				Umm el-Jimal?
[Damascus	23		Damascus			
Aenos	37					
Chanata	20		Canata			Qanawat
Rhosa]	40					?
Hatita	11	Adittha 37.30,31	Anitha			Kh. Samra'
Gadda	13	Gadda 37.20				Hadid
Philadelphia	62		Philadelfia	Philadelphia		Amman
Rababatora	48	Areopolis 37.17		Areopolis		Rabba
Thornia	?		Thana		Thawa	Thawana
Negla	22		Necla			'Ain Nejl
Petris	18		Petra	Petra	Petra (?)	Petra
Zadagatta	20	Zodocatha 34.24	Zanaatha			Sadaqa
Hauarra	24	Havarra 34.25	Auara			Humeima
Praesidio	21	Praesidium 34.35			Praesidium	Kh. el-Khalde
Haila		Aila 34.30	Elana	Aila		? 'Aqaba
		Zoara 34.26	Zoara			
		Arieldela 34.44				Gharandel
		Admetha 37.18				W. Itm?
		Ziza 37.16	Ziza			Zoza, e. of Madeba
		Mafa 37.19				Umm er-Rasas
		Betthoro 37.22				Lejun?
		Libona 37.27				Kh. Libb
		Naarsafari 37.28				
		Asebaia 37.32				Kal'at ed-Daba'a?
		Ualtha 37.33				W. el-Walid?
		Vade Afaris fluvii in castris arnonensibus 37.34				
		Apud Arnona 37.35				R. Arnon

as Pella, Jerash, Madeba, Hesbān, and Kerak. The Peutinger Table draws on imperial administration, and Eusebius on biblical and early Christian writers and on his personal knowledge of the country. The author of the Peutinger map could not have known the region first-hand; for example, he makes a serious geographical error by having the River Heromicas (the modern R. Jarmuk) flow into the north-east corner of the Lacus Asphaltidis (Dead Sea) rather than into the Jordan a little south of the Sea of Tiberias. Eusebius, and the author of the Madeba Map (see below), show much more detailed knowledge of the region.

First millennium travellers to Palestine from western Europe (see Wilkinson, 1977) were mainly Christian pilgrims and scholars, such as Paula (AD 382), Theodosius (*c.* 518), Antoninus Martyr (*c.* 560–70), the Piacenza Pilgrim (*c.* 570), Arculf/Adomnan (*c.* 670), and St Willibald (*c.* 754). The Piacenza Pilgrim reveals knowledge of Antoninus who 'preceded' him (perhaps in a spiritual sense; cf. Wilkinson 1977, 79, n. 2). They have left no maps, but some of them show at least some knowledge of Transjordan, as the following table shows:

Theodosius	*Antoninus M.*	*Piacenza P.*	*Arculf*	*Willibald*
Bethsaida				Bethsaida Corozain
Jor & Dan	Jor & Dan	Jor & Dan	Jor & Dan	Jor & Dan
	Sea of Galilee	Sea of Tiberias	Sea of Galilee	
	Gadara	Gadara		
	Baths of Helias	Baths of Elijah		
Scythopolis	Scythopolis	Scythopolis		
		Livias		
	Salamaida (=Kefrein?)			
	hot baths	hot springs		
	Dead Sea	Salt Sea		
	Segor	Segor	Zoar	
		Sodom	Sodom	
		Gomorra		
	Sinai			
	Abila [= Ailath]			

These lists, especially those of Antoninus Martyr and the Piacenza Pilgrim, reveal both biblical and genuinely topographical knowledge. Roughly contemporary with them is the Madeba Map, which, though incomplete, reveals a little more of Transjordan in the Byzantine period, naming and locating the following places (see Pl. 1B): Aenon, now Sapsaphas (W. el-Harrār); [Ba]aru (Hammāmāt ez-Zerqā); Therma Kalliroes, the hot springs of Callirhoe (Ain ez-Zāra); [Char]achmoba (Kerak); Bethomarsea also Maiumas (Ain Sāra, below Kerak); Aia (Kh. 'Ay, 6 km s.w. of Kerak); Tharais (al-'Irāq, 5 km south of Kh. 'Ay); the river Zared; [the S]alt Sea also Asphalt Lake, [or De]ad Sea; Bela, also Z[oar, now] Zoora; the Deser[t]; to tou agiou L[ot], the sanctuary of Lot; Praesidium (for these identifications, see Donner, 1992). If Donner is correct in identifying three letters MEL(?) as part of a place-name [Ge]mel[a], the map may once have included reference to Petra and Edom (Donner 1982, 188–91; 1992, 41), but though this is possible it is hardly certain from this limited evidence. Donner (as other scholars) notes (1992, 22) that the main literary source for this map was Eusebius's *Onomasticon* (*c.* AD 320, translated into Latin by Jerome *c.* AD 400); '61 of altogether 149 inscriptions of the present fragmentary map are derived directly or indirectly from Eusebius' *Onomasticon*'. Of the names above, however, only the Zared, the Dead Sea, Bela/Zoar, and Petra, if admissible, appear to come from Eusebius. Praesidium appears as a Roman station on the route north from Aila to Petra in the *Notitia Dignitatum* and the Peutinger Table.

As is well known, the sixth-century Madeba Map gives pictorial representations of many of the places named. Thus in Transjordan the map provides illustrations for Calliroe (a bath), Karachmoba (a walled town), Betomarsea/Maiumas (a domed building with side-wings), Aia, Tharais and Zoara (fortresses), and the sanctuary of Lot (a church). Such illustrations appear on other sixth–eighth century mosaics in Jordan (see Piccirillo 1992), notably at the Acropolis Church at Ma'in (built AD 719/20) and at the Church of St Stephen at Umm al-Rasas (Kastron Mepha'a) (AD 756). At Ma'in the church mosaic is bordered with illustrations of named cities; those remaining on the north border are [Charach M]oba, Areopolis, Gadaron (probably the Gadara in Petraea), Esboun[ta](Hesbān), Belemounta (Ma'in). At Umm al-Rasas, in the intercolumnar spaces of the southern aisle aisle, appear seven cities from the Transjordan (Pl. 2): in sequence from east to west they are Kastron Mepha'a (Umm al-Rasas), Philadelphia, Medaba, Esbounta (Hesbān), Belemounta (Ma'in), Areopolis (Rabba), Charach Moba (Kerak). This list, apart from Kastron Mepha'a which has been placed at the top clearly from local pride (Pl. 3), runs from north to south, from Philadelphia to Kerak. At the head of one aisle appears the village of Limbon (Libb), and at the head of the other Diblaton (biblical Diblathaim). These mosaics do not exactly form a map, but they do indicate that their creators were aware of the wider geographical picture and the topographical relationships between the towns named. They are contemporary evidence for eighth-century Transjordan and its more important centres of habitation (see recently Bowersock 2006, 73–76).

We have therefore very little evidence for the pre-medieval mapping of Transjordan. The *Tabula*

Peutingeriana shows important stations on the main Roman roads through the region; its information is supplemented by evidence for military and civic centres from the *Notitia Dignitatum*. The Madeba Map draws largely on biblical tradition, mediated via Eusebius, and adds only a few names. Through a millennium of political, military, ecclesiastical and other human activity, it would be surprising if no other maps of the Transjordan or parts of it had been drawn — the mosaics from Ma'in and Umm al-Rasas (Piccirillo 1992) show an awareness of the towns of the region and their inter-relatedness — but no other traces of such local cartographic activity have been preserved to us.

CHAPTER TWO

MEDIEVAL SCHOLARS, TRAVELLERS AND MAPMAKERS

THE NEXT HALF-MILLENNIUM evidences a great increase in known cartographic activity for the regions of Palestine and Transjordan, especially in the twelfth century. This can hardly be unconnected with the activity of the Crusades and the establishment of the Latin kingdom of Jerusalem, and the development of pilgrimage. Mapmakers were aware of the Crusaders' castles at Kerak and Shobek (Mont Regalis/Montreal), while pilgrims were interested in the burial places of Job and Moses, and various other sites associated with events narrated in the Old Testament. Biblical scholars were exercised by the biblical evidence for the location of various places — for example, Idumaea, Seir, and Petra. The ground was well prepared for the cartographers by the travellers of the twelfth century and the study of the Dominican Burchard, to whom most subsequent writers and mapmakers were heavily indebted.

A BIBLICAL MAP FROM THE NINTH CENTURY

A ninth-century manuscript of a Latin commentary on the Bible held by the Bibliothèque nationale de France (BNF lat. 11561) contains a map illustrating the tribal geography of the book of Joshua. This map is probably a scribal copy of an earlier map (the scribe was clearly copying something he did not completely understand), and it appears to be an integral part of the commentary, whose original author was possibly an eighth-century Irish monk (for fuller detail, see O'Loughlin 2005). This map is titled 'Figura terre repromissionis', a phrase taken from the Vulgate of Heb. 11.9, presenting the land as 'the land of [God's] promise'. The author presents the land as systematically divided between the eleven tribes of the Old Testament (the twelfth tribe, the Levites, had no land). The River Jordan divides the map horizontally from north to south. At the very top a small inverted triangle suggests the division of the upper reaches of the Jordan into two parts (indicating the two springs, Jor and Dan, which from Jerome's time were seen as the two sources of the Jordan [Jerome, *Comm. In Evang. Mat.* at Mt 16.13]). The river flows south through a sea which is presumably the Sea of Chinnereth (Josh. 12.3) and eventually widens into a sea oriented west-south-west named as the *mare mortuum* (Dead Sea); from this a river, probably to be identified as the River of Egypt and as the southern border of Jewish territory (the Vulgate's *torrens Aegypti*, Josh. 15.4, 47), flows west into the Mediterranean. East of the Jordan, the map divides the land roughly equally, from north to south, between the half tribe of Manasseh, the tribe of Gad, and the tribe of Reuben (cf. Josh. 13.8-21). Each tribal area is allocated its city of refuge (cf. Josh. 20.1-6); thus Golan appears in half-Manasseh, Ramoth Gilead appears in Gad, and Bezer in Reuben (cf. Josh. 20.8). (The copyist has in error turned Ramoth Gilead into two separate places.) The problems of asylum were important to ecclesiastical lawyers in late seventh-century Ireland, and this perhaps explains the focus in this map on the cities of refuge, to the exclusion of all other sites in Transjordan mentioned in the book of Joshua (cf. O'Loughlin 2005, 17). However that may be, it is clear that this map relies on biblical scholarship rather than first-hand acquaintance with the territory, and that its concern is to clarify the book of Joshua's presentation of tribal territories for the reader.

AL-IDRISI

Apart from the maps perhaps originally accompanying Ptolemy's *Geography* (their existence is disputed) and the Madeba Map (which was unknown from the end of the Byzantine period to the end of the nineteenth century), the first extant map showing Palestine and Transjordan in some detail in anything like a modern form is Idrisi's map of AD 1154 (G. LeStrange 1890; Miller 1987; Nebenzahl 1986, 34–35). Idrisi (*c.* 1099–1164) shows the coastal region of the Levant (Pl. 4) separated from the inland region by the long line of *jebel el-lekam* (*jebel libnan*); east of this Idrisi names (in Arabic) *ba'albek*, *dimask*, *adra'at*, the sea of Tiberias (*buhaire tabaria*), *'Amata*, *zoara*, the Dead Sea (*al buhaire el maita*), the river Jordan (*al ardan*), *kasr ibrahim*, *faran*, *aila*, and *madjan*, these last two on the coast of the Red Sea. A small promontory, almost certainly the Sinaitic promontory, divides *faran* and *aila*. (The tenth-century geographer Mukaddasi knew that Waila [Aila] stood on one arm of the China, i.e., the Red, Sea [Le Strange 1890, 549]; knowledge of the two gulfs at the northern end of the Red Sea goes back at least to Pliny [*N.H.* V.65], who names them as the *Heroopolites sinus* and *Aelaniticus sinus*.) The Sea of Tiberias and the Dead Sea appear, linked by the Jordan, east of the line of mountains separating the Mediterranean coast from the interior, but a spur of the Jordan turns off west over the mountains towards Gaza. Two mountain chains run off north-east from the north-south chain, one from the Sea of Galilee and the other south of the Dead Sea. Idrisi worked and produced his map at the court of the Norman ruler Roger II in Palermo in Sicily, and his work was known to the fourth-century mapmaker Petrus Vesconte, but there is little sign that his representation of Transjordan was known and used by western cartographers of the Holy Land.

TWELFTH-CENTURY EUROPEAN TRAVELLERS, THEIR DESCRIPTIONS AND MAPS

Quite independently of al-Idrisi, and probably in complete ignorance of him and his work, several Europeans wrote accounts of the Holy Land in the twelfth century, including in their work a certain amount of information about the Transjordan. From the first half of the twelfth century we have a *Descriptio locorum circa Ierusalem adiacentium* from a manuscript (Fonds Latin no. 5129) in the Bibliothèque Nationale, Paris, edited by M. de Vogüé (de Vogüé 1860: Appendice, 407–33). This manuscript was written *c.* 1151–57, but de Vogüé argued that the basic *Descriptio* was probably earlier, *c.* 1130. This was at some stage credited to a Fretellus Archidiaconus, and with additional material was published under the title *Liber locorum sanctorum Terrae Jerusalem* in *Stephani Beluzii Miscellanea*, tom. I, 435–39 (Lucca, 1761). Another quite different version, entitled *Tractatus de distantiis locorum Terrae Sanctae*, and credited to Eugesippus, AD MLX, appears in Leo Allatius' *Symmicta* (Col. Agripp. 1653, pp. 114–20), republished in J. Genesius' *De rebus Constantinopolitanis* (Venice, 1733). An English version, translated by J. R. S. Macpherson, is published in Palestine Pilgrims' Text Society v, pp. 8–54 (London, 1892) under the name of 'Fetellus'. P. C. Boeren edits a *Descripcio cuiusdam de locis sanctis* under the title *Rorgo Fretellus de Nazareth* (Amsterdam, 1978). John Wilkinson (1988, 12–16) disentangles the evidence and presents a clear stemma, beginning with 'The Work on Geography (*Wge*)' (1128–37), which was the basis for the *Descriptio* (1131–43) and Fretellus' letter to Bishop Henri Sdyck (*Patri H.*) in 1137. This latter was used as a source, first by Fretellus for another letter (*Fratri R.*, a version of which appears in the *Tractatus* listed above), and then by John of Wurzburg and Theoderich later in the century. However, the contents of this *Descriptio* and its references to the Transjordan are extremely interesting, and are closely comparable with those of John of Wurzburg's *Descriptio* (*c.* 1160), Theoderich's *Libellus de locis sanctis* (*c.* 1171–73) and William of Tyre's *Chronicon* (*c.* 1180). These works mention more or less the same places in Transjordan, and seem to draw on the *Descriptio* or at least to share a common source. The following table (presenting the names in a roughly north-south sequence) will reveal some of the similarities, and give a picture of twelfth-century Transjordan as known to western European scholars of the time.

Descriptio	*J. Wurzburg*	*Theoderich*	*William of Tyre*	*Burchard*
c. 1130	*c.* 1160	*c.* 1171–73	*c.* 1180	*c.* 1280
Syria	Syria	Syria	Syria	Syria
Damascus	Damascus (head of Syria)	Damascus	Damascus	Damascus
Idumaea		Idumaea	Idumaea	Idumea
Edom		Edom	Edom	Edom
Seir, Sanyr		Seir		Seyr, Mons Senir
Traconitis		Traconitis	Traconitis	Traconitis
Ituraea		Ituraea	Ituraea	Ituraea
Arabia	Arabia	Arabia	Arabia (Prima, Secunda, Tercia)	Arabia
Bostron/Bosor	Bostron	Bosra	Bostron/Bosseret	Bostrum/Busere
Malbech (Baalbek)			Malbech/Eliopolis	
Paneas/Balinas/ Caesarea Philippi	Paneas/Belinas	Paneas/Belinas/ Caesarea Philippi	Baneas/Belinas/ Caesarea Philippi	Balinas
Meddan	Meddan plain	Medan plain	Medan	Medan
Suetha/land of Hus/ monument of Job	Suata/land of Hus/Job	Sueta/land of Hus/Job	Suita	Suetha
Cedar	Cedar/Gadara	Cedar	Gadara	[Gamela]
	Plain of Thorns	Plain of Thorns		
Ior & Dan	Ior & Dan	Ior & Dan	Ior & Dan	
Bethsaida	Bethsaida	Bethsaida		
Chorazin	Chorazin	Chorazin		Chorazin
				Gaulon
				Ramoth Gilead
Aulon/Gorius	Aulon/Gorius	Great Valley/ Valley of Meadows		
			Decapolis	
Gergersa	Gerasa		Gerasa	
			Yppus	
			Pella	Pella
			Mons Galaat	Mons Galaad
			Iazara	
Iaboc	Iaboc (river)	Iaboc		Iaboc
			Philadelphia	
Baal &Belmon (Reuben)				
Betharam (Gad)				
			Syra Sobal	Syria Sobal
Lacus Asfaltites	Lacus Asfaltis	Lacus Asfaltites	Lacus Asfaltis	Lacus Asfaltis
Mare mortuum	Mare mortuum		Mare mortuum	Mare mortuum
Segor/Bala/Zoara/ Casale Palma	Segor/Bela/Zora/ Palmaria/Palma	Segor/Bala/Cara	Segor	Segor
				Mt Nebo
Sodoma		Sodoma	Sodoma	Sodom
Gomorra		Gomorra		Gomorra
Seboim		Seboim		
Adama		Adama		
Medaba				
Mt Karnaim on Mts of Moab	Karnaim (Kerak)	Carnaim on Mts of Moab		
				Petra/Rabbath/ Areopolis=Ar in torrente Arnon cap.of Arab.Sec.
alia Bozor in montibus Idumaeae de qua Isaias				
	vallis Moysi	vallis Moysi	vallis Moysi	vallis Moysi

Descriptio	*J. Wurzburg*	*Theoderich*	*William of Tyre*	*Burchard*
Helim	Helim		Helim	
Horeth	Mt Horeb			
Mons Abarim	Mons Abarim	Mons Abarim		Mons Abarim
	Montreal	Mons regalis	Mons regalis	Montreal also
			Chrac, postea	called Krach
			Petra Deserti	olim PetraDeserti
Synai	Sinai	Sinai		
				Seir/Edom
				over against
				wilderness
Mare rubrum	rubrum mare	rubrum mare	rubrum mare	of Red Sea

One notes immediately that the names for the northern regions are drawn mainly from Roman sources, and that those for the southern region mainly from the Bible, with the obvious exceptions of Lacus Asfaltis (from Josephus) and the Crusading castle Mons regalis (Shobek). The *Descriptio* included also the list of Israelite halting places — *mansiones* — from Num. 33 (see de Vogüé 1860, 418–20; Wilkinson 1988, 185), whose final names — Asion-gaber, Desertum Sin, Mons Or, Selmona, Fynon, Hebar, Oboth, Dibungat, Selmon Deblataim, Mons Abarim and Campestria Moab — contributed to the twelfth-century picture of southern Transjordan. One notes also that these authors create some confusion for themselves and their readers by locating together Arabia, Idumaea, Seir and Edom, and that Idumaea has become associated with Syria, bringing Edom and Mount Seir with it. Thus the *Descriptio* notes that

Arabia jungitur Idumea in confiniis Bostron, qua est Bosor ex qua Barach Buzitus. Est autem alia Bosor in montibus Idumee, de qua Isaias . . . Idumea partes sunt Trachonitis et Ituraea quasi ad Damascum respecientes . . . Bostron olim Idumae metropolis. Idumea sub Syria; in Syria Damascus . . . Damascus in Syria reverenda quondam metropolis. . . . Partes cuius habitavit Esau, qui et Seir et Edom. Ab Edom pars Syrie Idumea vocatur. In Seyr Idumea civitas. In Idumea haud procul a Damasco mons Seir (de Vogüé 1860, 420).

Arabia joins Idumaea in the territory of Bostra, which is *Bozrah*, from which came Barach the Buzite. But there is *another Bozrah in the mountains of Idumaea*, about which Isaiah [said] . . . *Trachonitis* and *Ituraea* are the parts of Idumaea that *look towards Damascus* . . . Bostra was once the capital of Idumaea. Idumaea is under Syria and in Syria is Damascus . . . Damascus was once the capital city of Syria . . . Parts of it were inhabited by Esau, who is also called Seir and Edom, and from Edom part of Syria is called Idumaea. In Idumaea not far from Damascus is Mount Seir. (Wilkinson 1988, 188–89)

Similarly, Theoderich says (ch. 45) that '[the Jordan] divides Galilee from Idumaea and the land of Bosra, which is the second capital of the Idumaeans, next to Damascus', and in ch. 49 that 'Arabia joins Idumaea in the district of Bosra. Idumaea is a province of Syria. Damascus is the chief city of the Idumaeans . . . In Damascus once lived Esau and Seir and Edom, after whom all that land is called Idumaea . . . In Idumaea is Mount Seir, upon which stands Damascus' (Stewart 1997, 65, 70). The source of this confusion, or at least a contributory factor, may be that the biblical Bozrah of Edom became confused with Bosra or Bostra of Traconitis. 'Fretellus' notes the other Bozrah in the mountains of Idumaea (Isa. 63.1); this reference in fact goes back to Jerome's *Liber locorum* and Eusebius' *Onomasticon*, which distinguish Bostra, the mother-city of Arabia, and Bosor, the city of Esau in the mountains of Idumaea mentioned by Isaiah (Klostermann 1904, 46, 47; Freeman-Grenville, Chapman & Taylor 2003, 32).

The twelfth-century view of Transjordan seen in these writers is largely confirmed by the twelfth-century map of the Holy Land from the Biblioteca Medicea Laurenziana in Florence (Ashburnham, *Libri* 1882; cf. Röhricht 1895, 177 & Tafel V) (Fig. 2). According to Harvey (2003, 55–56), 'the map draws heavily from some text belonging to the closely

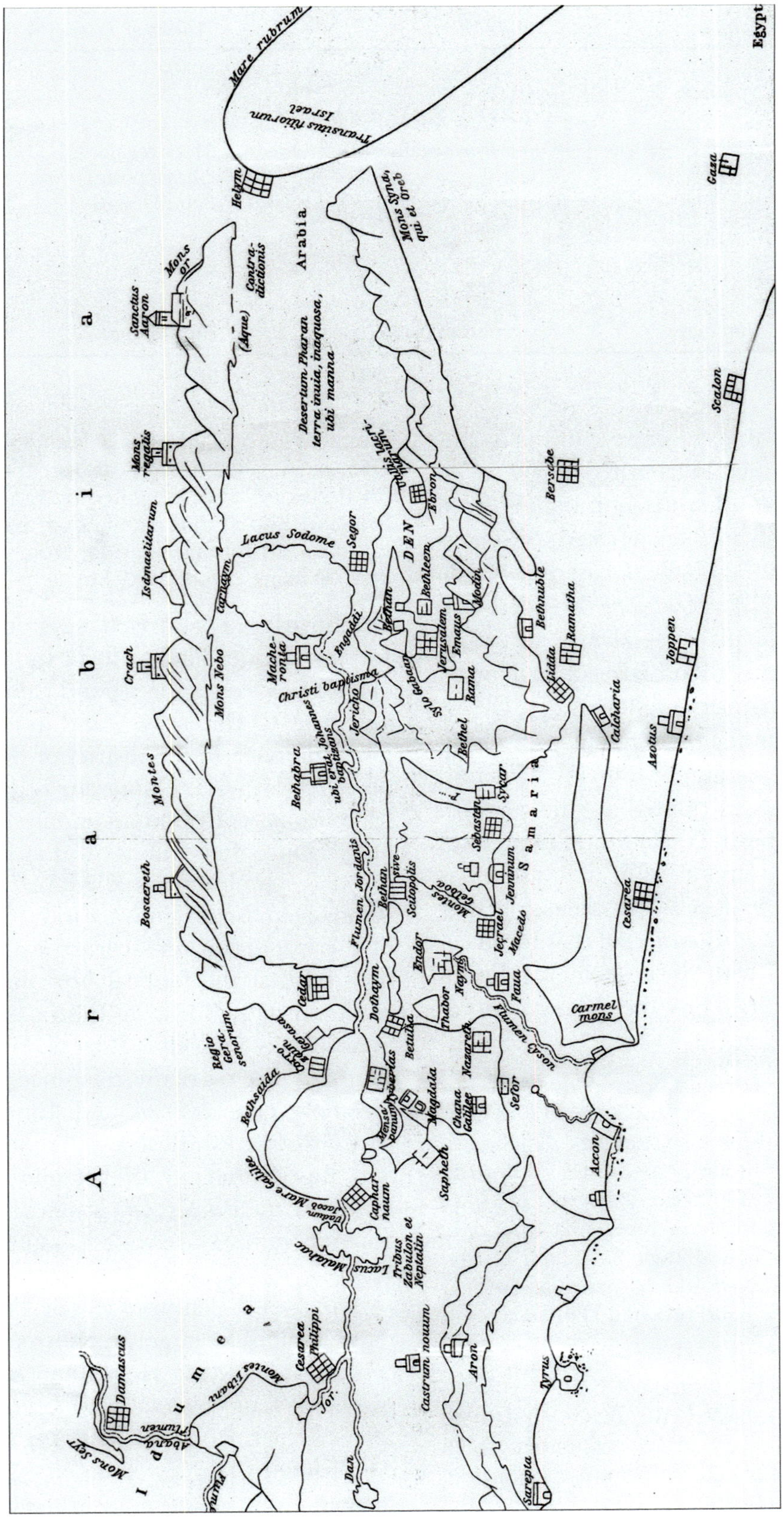

FIGURE 2. A twelfth-century map of the Holy Land from the Biblioteca Laurenziana, Florence (Röhricht 1895, Tafel V).

related group of descriptions that includes those by Fretellus and Theodericus and the so called Sixth Anonymous Guide, all of the mid or late twelfth century'. This map presents Arabia at the top and the Mediterranean Sea at the bottom, and east of the Jordan valley begins with Syria to the left, followed by Mons Seyr, with Damascus immediately beneath it, in the region of Idumaea. The Meddan plain, with the two sources of the Jordan, Dan and Ior, with Caesarea Philippi, are located nearby to the west of Damascus. Then follows to the south the Sea of Galilee (with Bethsaida and Corrozaim on the east side), Gerasa and the Regio Gerasenorum, Cedar (= Gadara), Bosaereth (= Bostra), the Montes Ismaelitarum with Crach (= Kerak) and Mons Regalis (Shobek). West of Crach, towards the Lacus Sodome, is shown Mt Nebo and, to the south of it, Carnaym, which is clearly close to the mountains of Moab but is not identical here with Crach/Kerak. The Jordan wriggles down from the Sea of Galilee (which is quite accurately shaped) to the Lacus Sodome (i.e., the Dead Sea, which is not accurately shaped); on the left, eastern bank of the river Jordan lie Bethabra ('ubi erat Johannes baptizans') and Macheronta (Machaerus, John's prison). East of the Lacus Sodome, the Montes Ismaelitarum continue south towards Sanctus Aaron and Mons Or, reaching the Mare rubrum (Red Sea) at Helym (Aila); south of the Lacus Sodome the map shows Segor, the Desertum Pharan, and the Aquae Contradictionis (a reference to the story of Num. 20.2-13). A spur of mountains running south-east from Judah runs out at Mons Syna 'qui et Oreb', near the Mare rubrum.

This clearly drawn map has many points of contact with the twelfth-century authors. In particular, it shows Damascus in close relationship to Mt Seir and Idumaea. It shares with them the names of Planicies Meddan, Dan and Ior, Caesarea Philippi, Bethsaida, Corrozaim, Gerasa, Cedar, Scitopolis/Bethan (locating these west of the Jordan), Bosaereth (Bostra), Mons regalis, Carnaym, Mons Or, Segor, Helym, Mons Syna/Oreb, and Mare rubrum. Bethsaida and Corrozaim are located east of the Sea of Galilee, with Capharnaum on the west side; similarly John of Wurzburg and Theoderich have the Jordan flowing into the Sea of Galilee between Bethsaida and Capharnaum, and set Chorozain six (John of Wurzburg) or four (Theoderich) miles from Bethsaida, apparently on the eastern side of the Lake (this location goes back at least as far as Willibald, *c.* 754; cf Wilkinson 1977, 127, map 38). Clearly the limited information given by the Bible was formative in the authors' accounts and in the construction of the map, but interest in geography is clear in the discussion of the origins of the Jordan river, and knowledge brought back by Crusaders has contributed to the inclusion of Crach and Mons Regalis.

THIRTEENTH-CENTURY MAPS; THE *DESCRIPTIO TERRAE SANCTAE* OF BURCHARD OF MT SION

The same traditions continue into the thirteenth century, though there is less evidence to hand. There is a map from the first half of the thirteenth century (made possibly between 1235 and 1242 [Edson 1994, 21], 'either in preparation for or as a reflection of the crusade of Richard of Cornwall' [Edson 1999, 121]), which belongs to Corpus Christi College, Oxford (MS 2*, now mounted separately but formerly fol. 2v of a bible) (see the drawing of this map in Röhricht 1895, 177, Tafel VI; Nebenzahl's illustration [1986, 39] regrettably does not show the right-hand page and the Transjordanian area). It was R. Vaughan who first identified this map as the work of the Benedictine monk Matthew Paris (*c.* 1200–59) (Vaughan 1958, 235–50). P. D. A. Harvey (2001, 165–67; 2003, 56) has recently argued that this map is a copy in Matthew Paris's hand of a map published *c.* 1230 to which he had only temporary access. This map (Fig. 3) in fact shows very little east of the Jordan. Damascus is described as 'civitas caput Syrie' (cf. Fretellus, 'Damascus in Syria reverenda quondam metropolis'), but there is no sign on the map as preserved of Mt Seir or Idumaea nearby. The usual Fons Ior and Fons Dan are shown, the two streams joining before entering the Mare Galilee; but Corozain and Bethsaida are shown clearly on the west side of the sea. East of the Sea of Galilee there is an unexplained entry 'De morte et sepultura marescalli'. Between the Sea of Galilee and the Mare Mortuum is shown [the place] 'ubi baptizatus est Dominus ecclesia (Johannis) baptiste', and close by it 'Terra Hus sive patria Iob et ubi civitas Zeb, Zebec, Salmana' (a reference

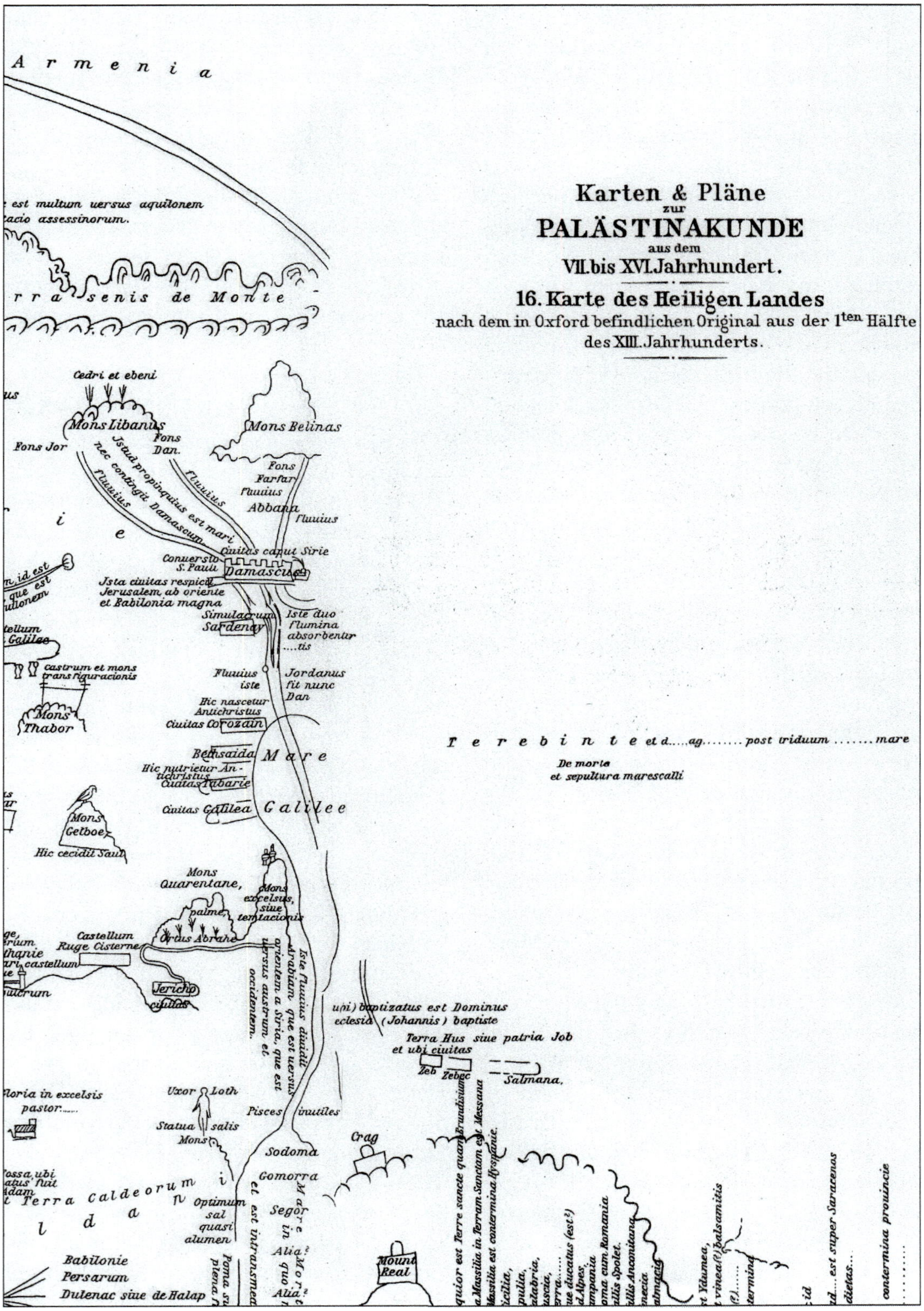

FIGURE 3. A thirteenth-century map from Corpus Christi College, Oxford, attributed to Matthew Paris (Röhricht 1895, Tafel VI).

to the Midianite kings of Judg. 7.25, 8.4-21). In the Dead Sea are shown Sodoma, Gomorra and Segor (perhaps Zoar; Harvey, id., 167, identifies it with Baale, i.e., Bela of Gen. 14.2); Admah and Zeboim are indicated by 'alia', 'alia'. East of the Dead Sea, from north to south, appear Crag and Mount Real (as on the twelfth-century map from Florence). A list of names at the bottom right of the map includes Idumaea among a group of names apparently referring to Arabia and including the Saraceni. This map reveals a very limited knowledge of Transjordan compared with the earlier map from Florence. Matthew Paris produced another map of Palestine, focused particularly on the fortified port of Acre, which appeared in different editions of his *Chronica Majora*. This map shows even less of the region east of the Jordan. It shows Damascus, the two sources of the Jordan, Ior and Dan, uniting in the River Jordan which flows directly into a square-shaped Dead Sea and apparently out of it at the other end, but nothing else (Lewis 1987, 347–62; Edson 1999, 121; Harvey 2003, 55–56).

The other piece of evidence for the thirteenth century is much more important. This is the *Descriptio Terrae Sanctae* by Burchard of Mount Sion (for a critical Latin text, see Laurent 1864; see also Grynaeus 1555, Canisius 1604; Eng. tr., Stewart 1896). Burchard, in Latin Brocardus, was a Dominican friar from Magdeburg or Strasburg, who probably wrote his *Descriptio c.* 1280–83. Burchard's *Descriptio* exists in a shorter and in a longer version, the former possibly written in the Holy Land and the latter after Burchard's return home (see Harvey 2003, 57). For his sources Burchard had the Bible, the *Onomasticon* of Eusebius, and the more recent itineraries of the twelfth century, including John of Wurzburg and Theoderich, the *Chronicon* of William of Tyre, and above all the *Historia Orientalis* of Jacques de Vitry (died 1240). Burchard's list of places in Transjordan is almost identical with that of his twelfth-century predecessors. In the north he added Gaulon and Ramoth Gilead. In the south he attempted to resolve a number of topographical problems, but succeeded only in complicating matters. He identified Montreal with Krak, which was strange because the two castles were well known, and compounded the problem by identifying the pair with the former Petra Deserti, here following William of Tyre who had identified Chrac (Kerak) with Petra Deserti (*Chronicon* 15.21; 20.26). He lists also a Petra which he identifies as Rabbath and as the capital of Arabia Secunda, identifying this with Areopolis (modern Rabba) and with Ar in the middle of the valley of the Arnon (cf. Dt. 2.16, 36) (Stewart 1896, 7, 58). It is worth quoting in full in this connection his paragraph on Arabia (*Locorum Terrae Sanctae exactissima Descriptio* in Grynaeus 1537, 296; cf. Laurent 1864, 22):

> Una est Arabia, quae Bastrum habet metropolim. . .Alia est Arabia Petraea, habens Petram metropolitem, quae olim Rabbath dicebatur, sita super torrentem Arnon. Hanc regionem olim inhabitarunt Ammonitae, tametsi Ar civitas fuerit Moabitarum: fuitque in ea regnum Sehon & regnum Og regis Basan. Tertia est Arabia, quae habet metropolitem Montem regalem, qui olim erat Petra Deserti, sita supra mare mortuum, continens sub se montem Moab, & vocatur proprie Syria Sobal, intra cuius terminos continetur Idumaea sive mons Seir.

> There is one Arabia, whose metropolis is Bostron. There is another Arabia, Arabia Petraea, whose metropolis is Petra, once called Rabbath, located above the river Arnon. This region was once inhabited by the Ammonites, although Ar had been a Moabite city, and there lay in it the kingdom of Sihon and the kingdom of Og, king of Basan. There is a third Arabia, whose metropolis is Mons Regalis [Montreal], which was once Petra Deserti, located above the Dead Sea and holding below it the mountain of Moab. It is properly called Syria Sobal, within whose borders is contained Idumaea or Mount Seir.

Thus Burchard thinks of three Arabias:

(1) Arabia prima; capital Bostrum;
(2) Arabia secunda, Petraea; capital Petra/Rabbath /Ar, above the Arnon (cf. Grynaeus 1537, 311: Areopolim . . . quae nunc Petra dicitur & est metropolis totius Arabiae secundae. Olim vero Ar vocata est, sita in torrente Arnon, in confinio Moabitarum, Ammonitarum & Amorreorum);
(3) Arabia tertia; capital Mons regalis, formerly Petra Deserti, above the Dead Sea; properly called Syria Sobal and containing Idumaea or Mount Seir.

and, confusedly and confusingly, of two Petras:

(1) Petra Deserti at Mons regalis in Arabia Tertia. William of Tyre had a century before identified Kerak with Petra Deserti (*Chronicon* 13.2; 15.21; cf. Laurent 1864, 43, 58; Stewart 1896, 38, n. 1);

(2) Petra, once called Rabbath and located above the Arnon in Arabia Secunda (Laurent 1864, 59). This second identification confuses Ar on the Arnon/Raba with Rabbath Ammon, as did his source William of Tyre (*Chronicon* 15.21).

This suggests that Burchard perhaps had in mind three Arabias, two of which had a Petra as its capital. Burchard also seems to identify three separate Idumaeas. Burchard locates one Idumaea in northern Transjordan, stating that the entrance to Idumaea is near Chorazin and the ascent of Mount Sanyr (Stewart 1896, 29). But Burchard also states clearly elsewhere (Stewart 1896, 8) that the third Arabia 'contains the land of Moab . . . and all ['all' is not in the Latin text] Idumaea', and then again (Stewart 1896, 38–39) that Idumaeans are to be found in Mount Seir in the country of Ashdod and Ascalon. Burchard is perhaps even more confused about Idumaea than his predecessors. He is certainly confused about the location of Petra; it was for a later Burckhardt to establish its real position.

Burchard also incorporates into his *Descriptio* a dissertation on Mount Seir. He notes the existence of three separate Mount Seirs:

(1) part of the kingdom of Basan, near the Sea of Galilee and Mt Gilead, also called Mount Sanir (Dt. 3.8,9) and Mount Hermon;
(2) Mount Seir or Edom over against the wilderness of the Red Sea (Gen. 14, Dt. 2) adjoining the wilderness of Paran, cf. *Novus orbis regionum*, Basileae 1537, 305: 'Est autem tota illa terra ultra mare Galilaeae montuosa, eratque olim de regno Regis Basan, continens montem Seir; nam & Seir & Hermon inter alia montana comprehenduntur. . . .Est & alius mons Seir iuxta desertum maris rubris Pharan');
(3) in the country of Ashdod and Ascalon, whose inhabitants (e.g., Antipater and Herod) were called Idumaeans (Stewart 1896, 34–38). This triplication of Seirs, near Damascus, near the wilderness of Paran, and near Ashdod and Ascalon, is obviously not unconnected with his presentation of Idumaea and the Idumaeans.

It is believed that Burchard produced a map to accompany his *Descriptio*, but no copy has survived. Röhricht has drawn attention to a map in the archives of Florence which he dates to the end of the thirteenth century, *c.* 1300, and so perhaps from the decades immediately after Burchard's work (cf. Röhricht 1891, 8–11, taf. 1; Harvey 2003, 57). This map (Figs 4 and 5) perhaps shows the influence of Burchard's views in that it presents an Ydumea regio next to Mons Hermon and Damascus, with a Mons Suanir, a Mons Seyr (but not identical with Mons Suanir) and a regio Traconitidis in close association with Ydumea; it locates another Mons Seyr further south, between Mons Galaad and the Jordan, and a third Mons Seyr at the southern end of the Dead Sea, south of Terra Edom. This is not precisely Burchard's view of Transjordan, but it is similar. This map shows Petra deserta, roughly in the position of Crac/Kerak, but shows no other Petra. Arabia is not mentioned. The course of the river Jordan and its tributaries anticipates that of the Sanuto-Vesconte map (thus, for example, the Arnon enters the Jordan well north of the Dead Sea), and the list of place-names displayed has more names in common with Sanuto-Vesconte's list than with Burchard's. The map divides the northern part of Transjordan between Neptalim, Ruben and Gad (listing from north to south) (with no mention of Half-Manasseh). This map is dated *c.* 1300, which is only a few years before the Sanuto-Vesconte map of 1321, and it is hard to believe that Sanuto-Vesconte had no knowledge of it. Nebenzahl (1986, 42) notes that 'In 1306 Sanuto drew the maps for the first version of his book, but they have been lost'. It seems not impossible that this map from the Florence archives is Sanuto's (or Vesconte's) first draft; P. D. A. Harvey (2003, 57) sees it as an early copy of Burchard's map.

THE FOURTEENTH CENTURY: PETRUS VESCONTE AND MARINO SANUTO

Marino Sanuto, a learned and well-travelled Venetian nobleman, presented a copy of his 1306 book, *Liber Secretorum Fidelium Crucis super Terrae Sanctae Recuperatione et Conservatione*, together with a set of maps drawn to Sanuto's instructions by the Genoan cartographer Petrus Vesconte, to Pope John XXII, on 24 September 1321. (The text is to be found in J. Bongarsius (ed.), *Orientalis Historiae Tomus II* [Hanoviae 1611].) Among the maps was a map of the Holy Land (Fig. 6); manuscript copies have survived at Brussels, Florence, London, Oxford and Rome (Röhricht 1898, 90–91). Sanuto's book, and the text

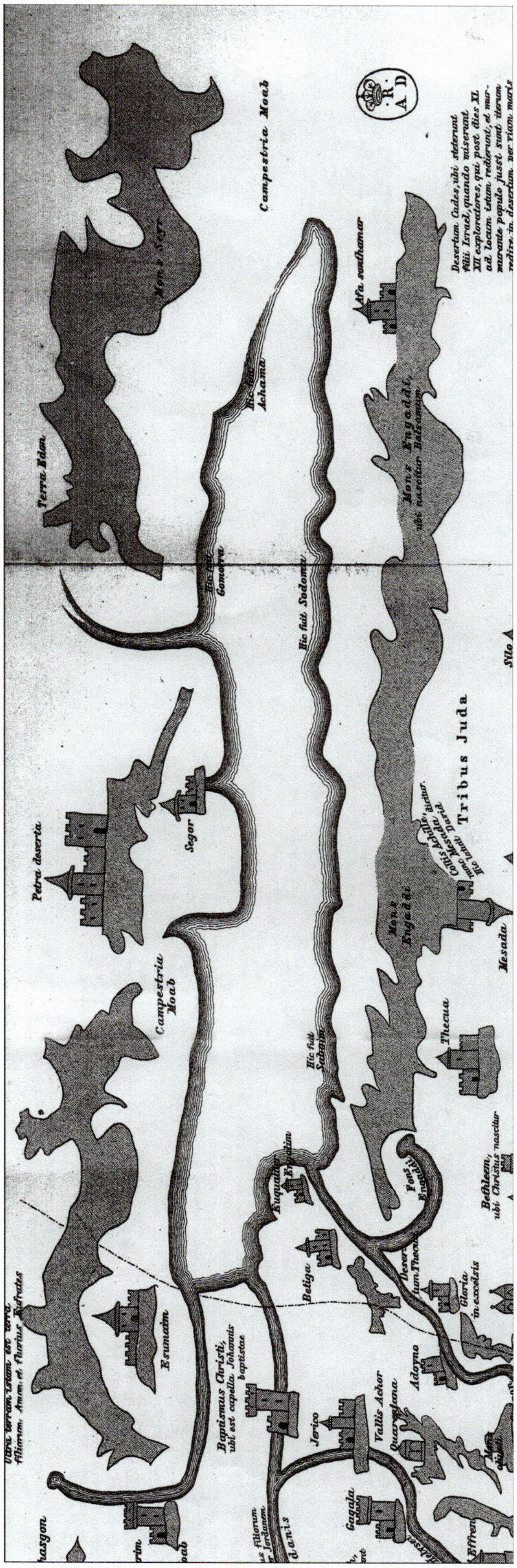

FIGURE 4. Map of the Holy Land from Florence, *c.* 1300 (southern half) (Röhricht 1891, Tafel I).

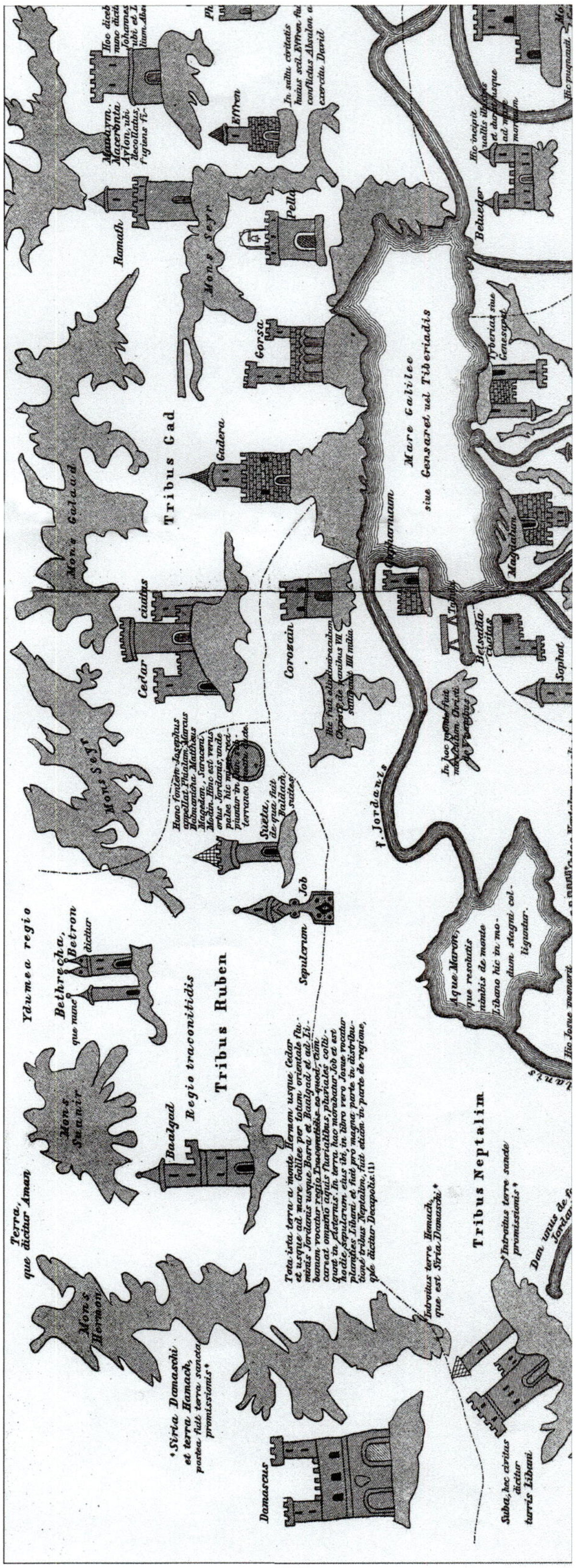

FIGURE 5. Map of the Holy Land from Florence, *c.* 1300 (northern half)(Röhricht 1891, Tafel I).

FIGURE 6. Map of the Holy Land, Petrus Vesconte and Marino Sanuto, 1321 (*Liber Secretorum Fidelium Crucis super Terrae Sanctae Recuperatione et Conservatione, apud Orientalis Historiae Tomum II*, ed. J. Bongarsius, Hanover 1611). (*The Board of Trinity College, Dublin*)

on Vesconte's map, draw heavily on Burchard (Röhricht 1898, 92–94). Thus Sanuto's map shows a mountain north-east of Lake Galilee with the legend 'Mons Sanyr <Seyr> coniungitur Damasco' (though Damascus is shown just north of Dan, well west of its true position), and just north of Mons Sanyr/Seyr it shows Bosra, regio Bostro, and Idumaea. A place named Seyr is shown imediately south of the torrens Arnon. A second Mons Seyr is shown alongside Terra Edom at the southern end of the map. In line with earlier tradition, Sanuto's map shows Ior and Dan as sources of the Jordan, the tomb of Job, Sueta, Medan, Cedar, Gadera, Gerasa, Pella, Galaad, Ar, Areopolis, Arnon, Mt Abarim, Nebo, Petra Deserta, and the vallis salinarum/lacus asfaltidis/mare sodomorum vel maledictum vel mortuum. However, Sanuto and Vesconte mark out the tribal territories of Naphtali (partly east of the Jordan valley), half Manasseh (which does not appear on the map from Florence), Ruben and Gad. The entrance to Hamath, Dan, and Corocaym [Chorazin] appear in Naphtali; Cedar, Gadera, and Gerasa in half Manasseh; Efron, Fanuel, Basan, Iabes, Iayr vel avoth Seyr, and Manay vel machero[n]ta nu[n]c haylon [Machaerus] in Gad; and regia Seon regis Esebon, Esebon, Kmon [Hermon], Eleale, Betsemut, Iacer [Jazer], Fagor [Peor], and torrens Arnon in Ruben. South of Ruben appear Ar, Fagor, Seyr, Abarim, Nebo, Fasga [Pisgah], Betonaym, Petra Deserti, Terra Edom and Mons Seyr. The region east of the Jordan is neatly defined by a range of mountains running from Traconitis or Terra Hus in the north to Terra Edom in the south.

This presentation of Transjordan, however, is full of inaccuracies, most of them anticipated by the Florence map. Damascus is shown too far west, the tribe of Naphtali extends east of the rift valley, Chorazin is east of the Sea of Galilee, the river Arnon empties into the Jordan well north of the Dead Sea (instead of entering the Dead Sea's eastern shore in the middle) and consequently Machaerus, Heshbon and Seyr are located too far north; Abarim, Nebo and Pisgah are shown south of the Arnon, and Petra Deserta is identified with both Crac and Mons Regalis (as in Burchard), though these two places, now Kerak and Shobek, are quite distinct. It is clear that Sanuto and Vesconte have no first-hand knowledge of the Transjordanian region, drawing largely on the biblical traditions via Burchard. A Crusader using this map would have become very confused.

Equally interesting, though less detailed, is the presentation of the Transjordan region on Sanuto-Vesconte's *Periplus of Asia Minor, Syria and Egypt*, from the same set of maps (cf. *Orientalis Historiae Tomus II*, map II, Hanover 1611; Röhricht 1898, Tafel. 19; Stewart 1897, ill. II). This map (Fig. 7) shows in great detail the towns of the Mediterranean coasts of southern Turkey, Syria, Lebanon, Israel/Palestine, and Egypt, and in much less detail their hinterland. The Jordan, with its two sources, the Waters of Merom, Sea of Galilee and the Dead Sea are shown roughly parallel to the coast. Behind them to the east is a single chain of mountains running south to the Sinai peninsula and ending in Mons Synay and Ecclesia S. Catharine, immediately south of which is a clearly marked Meccha. This lies on the Arabian peninsula, with the Persian Gulf to the east, fed by the rivers Euphrates and Tigris, and the Red Sea to the west. The map does not portray the bifurcation of the Red Sea into the gulfs of Suez and 'Aqaba (unless Vesconte has confused the Gulf of 'Aqaba and the Persian Gulf). However, west of the mountain chain a little north of the Dead Sea is marked Petra, and south of the Dead Sea appear Ydumaea and Mons Seyr, with a rectangle inscribed 'Lex', presumably indicating that the Law was given here, and an indication of the Aquae marat (waters of bitterness, Ex. 15.23). East of the mountains is written, along the mountain chain from north to south, 'alia Syria quae Arabia tertia', 'alia Syria quae Arabia Secunda', and 'alia Syria quae Arabia prima', and finally 'Sobal et edemon [= *eudaimon*] vel sacra'. Sobal appears in Burchard as Syria Sobal, with reference to Moab and to Arabia Tertia, above the Dead Sea, with Mons Regalis as its capital. For Burchard, Arabia Prima has Bostron, further north, as its capital. Sanuto sees Arabia Prima as in the south; it includes the city of Mecca (Stewart 1897, 3–4). Burchard and Sanuto-Vesconte thus seem in some disagreement over the north-south sequence of Arabia Prima, Secunda and Tertia.

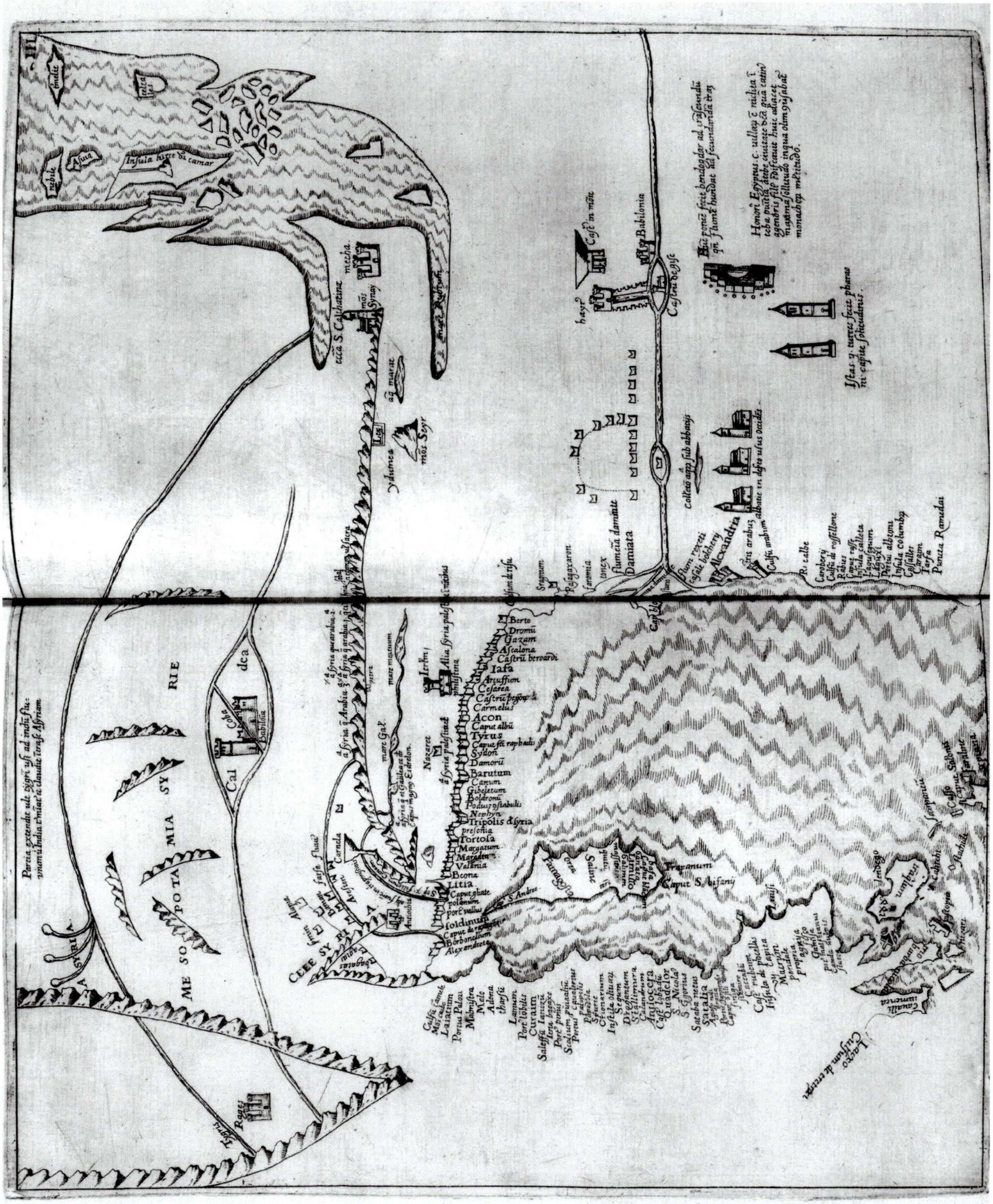

FIGURE 7. Periplus of Asia Minor, Syria and Egypt, Vesconte and Sanuto, 1321 (*Liber Secretorum Fidelium Crucis*, Hanover 1611). (*The Board of Trinity College, Dublin*)

OTHER FOURTEENTH-CENTURY TRAVELLERS

Two other travellers have left accounts of the Holy Land in fourteenth century with some reference to the Transjordanian region. The library in Trinity College Dublin contains the manuscript of a traveller from Dublin *c.* 1350; the manuscript was edited and translated by J. H. Bernard (1894). This traveller clearly knew the tradition, for he speaks of Bozrah where the boundaries of Arabia and Idumaea meet, and describes Idumaea as 'the land of Damascus'. In the south, he knows Lacus Asfaltis, Zoar, now the Town of Palms, notes Carnaim (Kerak?) 'as you go down to Arabia', and Mount Sinai in Arabia. From the same period comes Ludolph von Suchem's *Description of the Holy Land and of the way thither*, written in the year AD 1330, translated by A. Stewart (1895). Von Suchem knows Damascus, Belynas/Caesarea Philippi/Paneas, the Ior and the Dan, the Sea of Galilee/Lake Tiberias, the Dead Sea and Montreal, the 'strongest castle in the world' with the village Sabab (= Shobek?) at its foot. These travellers have little extra to offer, and perhaps confirm the suspicion that the Transjordan was less well known to western Europeans in the fourteenth century than in the thirteenth. Western acquaintance with the Transjordan was probably greatest in the twelfth century, when the great castles were built — Montreal (Shobek) in 1115, Le Vaux Moise (Al-Wu'eira) and Aila ('Aqaba) in 1116, and Kerak in 1142.

THE *MAPPAE MUNDI* TRADITION

The Latin term *mappaemundi* generally refers to maps of the world made in the medieval period (roughly from the fifth to the fifteenth century). They reflect a Christian world, in which the supreme authority is God, and find their authoritative source of knowledge in the Bible, with some assistance from the classical authors. The maps usually portray Jerusalem as the centre of the world; Rome, Constantinople and Alexandria are also important. The construction of such maps was 'a literary and theological exercise, not a geographical one' (Whitfield 1994, 14) and they were the result of imagination rather than empirical enquiry, for which the authors had little opportunity. For these reasons, however, the Holy Land and its surrounding nations are well represented at the heart of these maps, and it is interesting to see how they present the Transjordanian regions.

If we limit ourselves to the best known maps of this type — the Beatus map (Spain, *c.* 776, copied in 1109 [Nebenzahl 1986, 18–19]), the Cotton map (early eleventh century, probably from Canterbury, England [Nebenzahl 1986, 30–31; Barber 2006, 4–8]), the Sawley map (late twelfth century, from Yorkshire, perhaps a copy of a map used at Durham [Harvey 1997, 33–42; Barber 2006, 10–13]), the Psalter map (mid-thirteenth century, possibly a copy of a map from Westminster [Miller 1895, III, 37–43, Tab. 3; Whitfield 1994, 18–19]), the Hereford map (*c.* 1285, drawn in Lincoln [Edson 1997, 139–44; Harvey 2002; Harvey (ed.) 2006; Barber 2006, 27–30]), the Ebstorf map (*c.* 1300, from Lüneberg, Germany [Miller 1896, v, 38–44; Barber 2006, 23–27]), and the Higden map (drawn *c.* 1350 to accompany Ranulf Higden's *Polychronicon*, Miller 1895, III, 94–108; Edson 1998, 126–31; Barber 2006, 32–35]) — we find some interesting correspondences and evidence of what these mapmakers considered important to include, as the following table illustrates:

Transjordanian toponyms on medieval world maps

JEROME	*BEATUS (St Sever)*	*COTTON*	*SAWLEY*	*PSALTER*	*EBSTORF*	*HEREFORD*	*HIGDEN*
Montes libani			Mons Libanni		Mons Libanus Basan region, Basan ci Astharoth	Mons Libanus	Mons Libani
antilibanus							
Dan	Dan		Fons Dan	Fl.Dan		Fons Dan	
Ior	Ior		Fons Ior	Ior		Fons Ior	

JEROME	*BEATUS (St Sever)*	*COTTON*	*SAWLEY*	*PSALTER*	*EBSTORF*	*HEREFORD*	*HIGDEN*
Cesarea philippi		Cesarea			Caesarea	Caesarea	
vel abilina		philippi			philippi	philippi	
betsaida				bethesda			
				corosaim			
Damascus	Damascus		Damascus		Damascus	Damascus	Damascus
						civitas	
	Bostra					Bosor quae et	
						Bosra	
					Seyr mons		
siria					Syria	Siria sobal	
Mons ermon					Mons Hermon		
Mons galaad		Mons galaad	Mons Galaad			Mons Galaauth	
decapolis			Decapolis regio			Decapolis regio	
Mare galilea	[unnamed]		Mare Galilea	Stagnum genosar		Mare Galilee	
				Thiberiadis			
					Galaon		
					Gerasa		
					Gadera		
iordanis	[unnamed]				Jordanis		Iordanis fl.
			Ruben				
	[unnamed]		Iaboc			Torrens Jaboch	
					Arnon torrens,	Fl.Arnon	
					& fluvius		
					Aroen vallis,		
					mons		
						Montes Arnon	
		gad	Gad				
					Ramoth		
moab		moabite				Moabite	Region moab
					Dibon		
					Beelphegor		
					vallis		
						Mons Abarim	
		Mons fasga				Mons Fasga	
						Mons seir	
amorrei			Amorrei				
Idumea regio sirie					Idumea		Ydumea
amalech							
Desertum	Desertum						
cades	ubi filii Israel						
	xl annos						
	erraverunt						
		Dimidia	Dimidia tr.				
		tribus	manasse				
		manasse					
		ruben	ruben				
Mare mortuum	[unnamed]		Mare	Mare	Mare	Mare	Mare
			mortuum	mortuum	mortuum	mortuum	mortuum
					Pentapolis		
					Bale		
					Mons Segor		
					Adama		
					Gomorra		
					Sodoma	Sodom	

JEROME	*BEATUS (St Sever)*	*COTTON*	*SAWLEY*	*PSALTER*	*EBSTORF*	*HEREFORD*	*HIGDEN*
			Arnon			Arnon fluv. et civitas	
			Silva piperis			Silvas pipereas	
Mons sinai	Mons Sinai ubi filii Israel legem acceperunt	Mons sina	m.Sina		Syna mons	Mons Sinay	
						Mons Oreb	
					Thoman vel Saba	Theman	
madian			madian		Madian	Madian	madiane
Arabia terra	Arabia	Arabica deserta	Arabia deserta		Arabia	Arabia deserta, Saba	
saba							
						Petra civitas Arabie	
Arabicus sinus	Sinus arabicus				Sinus Arabicus		
	Nabatea regio						
Pharan							
Aila							

The first column of this table lists Transjordanian places appearing on the twelfth-century copy of the map of Palestine attributed to Jerome (*c.* 385) (Miller 1895, III, 13–19; Nebenzahl 1986, 18–19). It will be observed that, for whatever reason, the names appearing on the 'Jerome' map correspond closely to the lists of names from the *mappaemundi.* Prominent on the lists are Mons Libanus, Dan and Ior (the two accepted sources of the Jordan), Caesarea Philippi, Damascus, Mons Galaad, the Decapolis region, Mare Galilea, the river Jordan, Mare Mortuum, Mons Sinai, Madian, Arabia deserta, Sinus Arabicus. Clearly the physical features of the seas, mountains and deserts are important, providing a framework; Mt Sinai and Caesarea Philippi are the places of the revelation of the Mosaic Law (Exod. 24) and of the messiahship of Jesus (Mark 8.27-30); the Sea of Galilee and the Decapolis were also associated with Jesus' ministry (Mark 7.31). Damascus and Gilead have important associations with Abraham (Gen. 15.2) and Jacob (Gen. 31.22-54), the Sinus Arabicus (Red Sea), Sinai, and Midian with Moses (Exod. 13–24), and Damascus and Arabia with St Paul (Gal. 1.17). The connection of Transjordanian places marked on these maps with important biblical events and persons is clear. However, the makers of these maps did have a fair idea of the basic geographical relationship of the mountains, seas and rivers, however they came by it, and it is sometimes surprising how much detail they managed to include. The Sawley and the Hereford maps seem to list almost the same toponyms, including the interesting 'silva piperis'/'silvas pipereas' ('pepper woods'?); both include the Arnon and Iaboc (Jabbok) rivers. The Hereford map distinguishes itself by naming Petra civitas Arabie, 'Petra, the city of Arabia', but locates it north of the Arnon (cf. Burchard, 'Areopolim . . . quae nunc Petra dicitur et est metropolis totius Arabiae secundae', above p. 13).

THE CATALAN ATLAS

The Catalan Atlas (for an excellent modern reproduction, see Grosjean 1978) is a set of 8 maps usually attributed to a Jewish cartographer named Abraham Cresques (1325–87), a native of Majorca in the province of Catalonia, and is dated to 1375. It may have been commissioned by or for the emperor Charles V; the catalogue of his library shows that it was present there by 1380 (on the early history of the map see Grosjean, and Campbell's review [1981, 115–16]). The map combines features of the medieval navigational charts (*portolani*), and the illustrated world maps (*mappae mundi*), with information from the Arab cartographic tradition and from the narrative of Marco Polo. The section showing the Transjordanian region is dominated by a large vignette of Damascus in the north, with the Saracenic flag flying above it. Just south of this appear the two sources (unnamed) of the Jordan, which flows through Lake Huleh ('Aquaron'), the Sea of Galilee ('Mar de Gallilea'), and the Dead Sea ('Mar Gamora'). Parallel with the River Jordan, to the east, runs a mountain chain with a sequence of names, in red and black ink alternately, showing *mont de libanon, mont Hermon, mont Sanir, mont de Gilahd, mont de Pisga, mont Abarim, mont de Nebo, mont de Ruben* (i.e, Lebanon, Hermon, Sanir, Gilead, Pisgah, Abarim, Nebo and Ruben). The name of Ruben, here applied to a mountain, seems the odd one out, and one wonders why the tribe of Ruben was thus singled out, for it is not biblically associated with a mountain. To the south of the Dead Sea appears a vignette of St Catherine's on Mount Sinai with the inscription 'Hic est corpus Catarine virginis'. Below this appears the inscription' Mont de Sinay en lo qual Déu dona la ley a Moysses' ('Mount Sinai, on which God gave the Law to Moses'). Immediately west of this appears the city of Ellim on the eastern coast of the northern end of the Red Sea. South of Ellim appears the city of Essiongaber. What is striking about the Red Sea is that it is drawn with two prongs at the northern end, a western gulf with the route of the Israelites across it from Egypt clearly marked, and the eastern gulf with Ellim at its northern end. This feature may be derived from the map of Idrisi, on which it also appears, and known also from the practical experience of seafarers brought back to Majorca. The map notes that 'This sea is called the Red Sea, and it is through this sea that the twelve tribes of Israel passed. And you must know that the water is not red, but the seabed is red. Through this sea pass most of the spices that come from India to Alexandria'. The note about the seabed appears later on a map by Petrus Rosselli in 1465 (Winter 1952, 6, fig. 5).

This map is clearly more interested in the sea coast and its harbours, and in the general geography of the region, than in the roads and towns of the interior. Apart from the sequence of mountains with their biblical names, the cartographer is not interested in identifying or locating the biblical cities, except for the two ports on the Red Sea. The three important places inland are Damascus, Jerusalem, and St Catherine's monastery on Mt Sinai, sites of particular interest to western Christian rulers, travellers and pilgrims.

THE FIFTEENTH CENTURY: JOHN POLONER AND WILLIAM WEY

In 1421–22, a century after Sanuto and Vesconte produced their description and map, a German called John Poloner made his pilgrimage to the Holy Land. Poloner mentions that he has drawn a map (unfortunately not extant) (and on it has painted one kingdom yellow to distinguish it from the others) in which 'the whole of the Holy Land is divided into eighty-three squares and into twenty-eight spaces breadthways, by drawing lines across it, over the lines from west to east' (Stewart 1894, 24). As Vesconte's map has a grid of exactly the same number and arangement of squares, one must presume that Poloner's map is directly or indirectly drawn from Vesconte's. Poloner locates by reference to the grid a number of sites in Transjordan:

> Now in the second [Vesconte 3rd] space and the twelfth [Vesconte 13th] square is Bozra in the country of Bostoron, mentioned in Isaiah lxii [*sic*]. The road through the city of Cedrar (Cedar), in square 23 [Vesconte 22], leads to Aran (Aram), . . . Josephus calls this city Camela, because the mount whereon it stands is formed in the shape of a camel. In square 24 (53) in the mountains to the eastward is Areopolis, which once was called Aror [Aroer], the capital of Arabia Secunda, being four days' journey from Petra in the desert . . . Upon this Petra is built an impregnable castle once (called) Pirach (Kirach), where the Soldan lays up the treasures of Arabia and Egypt [cf. Vesconte's map]. Mid-way between this Petra and Areopolis there is the brook Sorec and the Mount Abarim . . . Three days' journey south from this Petra is Mount Seir, which borders on the wilderness of Pharan, and it is called the land of Arabia, even to the Red Sea. Note that at the edge of the mountains of Arabia Prima, round about Mount Sanir, begins the land of Huz, which is also called the province of Trachonitis, and extends even to Cedar and the Sea of Galilee, being partly formed by the country of Decapolis. Likewise from the edge of the

mountains of Arabia Secunda, even to the Jordan, was the kingdom of Og, King of Basan, which fell to the lot of the tribe of Gad as far as the brook Jabbok. I have painted this kingdom yellow to distinguish it from the others ... Between the brooks Ainan (Arnon) and Sorec is the plain country of Moab ... But the yellow country beyond the brook Sorec to the south is called the land of Moab and of Ammon, as aforesaid. Note that there are three cities of refuge beyond Jordan. The first, under square 23, near the hill country of Arabia, is called Golan [Josh. 21.27]. the second, under square 37, is called Ramoth Gilead [Josh. 21.38]; and the third, under square 33, is called Effrem, where Christ abode with his disciples [Jn 11.54]. (Stewart 1894, 25–26)

There are a number of points worth noting here. First, Poloner clearly places Arabia Prima in the region of Mt Sanir and Trachonitis, with Arabia Secunda immediately south of it, thus following Burchard and avoiding the error of Vesconte's *Periplus* map. Areopolis, as in Burchard, is in Arabia Secunda; Petra Deserti, as in William of Tyre, is identified with Kerak. The brook Sorec is presumably a version of the Zared or Zered; on Vesconte's map the Arnon enters the Jordan well north of the Dead Sea, and the next brook south (unnamed: = Poloner's Sorec?) enters the Dead Sea at its northern end. The location of the Ammonites and Moabites in this region south of the Sorec is therefore not unintelligible. Poloner includes three cities of refuge from the Transjordan: Golan, Ramoth Gilead and Effrem (Ephraim), but only the first two are found in Josh. 21 in the list of cities of refuge; the third comes from John's gospel (John 11.54). The two great castles of the south are almost forgotten; Poloner plays down Kerak, 'the impregnable castle once called Pirach' where the Sultan keeps his treasures, and does not mention Mons Regalis, Montreal, at Shobek, since 1293 in Mamluk hands.

In May 1458 William Wey, a Fellow of Eton College (founded 1440–41 by Henry VI), travelled to the Holy Land on pilgrimage, via Venice, from which he set sail across the Mediterranean on 18 May. Wey reached Jaffa on 21 June and Jerusalem on 24 June. His route through the Holy Land was limited and conventional, and included Rama, Lydda, Emmaus, Anathoth, Jerusalem, Bethany, and the River Jordan. He left Jerusalem on 2 July and was back in Venice by 6 September. He made a second visit in 1462, reaching Jaffa on 16 July and Jerusalem on the 19th; he left Jerusalem on the 26th and was back aboard ship on the 28th. He therefore spent very little time in the Holy Land itself, and could gain little personal knowledge of its geography. He almost certainly knew little or nothing of the Transjordan. His *Itinerary* was edited, with an Introduction, for the Roxburghe Club in 1857. In 1867 the Roxburghe Club published the *Map of the Holy Land illustrating the Itineraries of William Wey*. The author (Revd G. Williams: see the *Map* volume, 1867, introduction) of the Introduction to the 1857 edition of the *Itinerary* says that

The heading [p. 128 of the 1857 volume] 'In tabula ista sequenti continentur omnia in mappa Terrae Sanctae' led to a comparison of the names in the table with those in the map, and convinced me that the map before me was the one referred to. Had any doubt remained, it would have been removed by the following chapter (p. 132), headed 'Nomine civitatum & in mappa mea de Terra Sancta', where, after an alphabetical list of names, which are all to be found in the map, there follows (p. 138) 'Distancie locorum in Terra Sancta', which is almost an exact transcript of a table, with the same heading, prefixed to the map itself (p. xvi).

The list of names in Wey's Itineraries and the map show close dependency on the names on the Vesconte-Sanuto map and, to a lesser degree, on Burchard's text. (Röhricht [1904, 191] suggests that Wey found his sources in the Franciscan Library at Jerusalem.) There seems to be no original contribution here to the fifteenth-century's topographical knowledge of the Transjordan region. Wey's map continues the association of Bosra with Idumaea and Mt Senyr in northern Transjordan. It locates Mons Seyr near Ramoth Galaad on the south-eastern shore of the Sea of Galilee, and Petra Deserta at Kerak. It shows some traditional knowledge of biblical events recorded for Transjordan — the sepulchrum Iob near Idumaea, the stay of the Israelites in the plains (campestria) of Moab, the burial of Moses below Mt Abarim, the place of the Israelite crossing of the Jordan, and the tribal territories of Manasseh beyond the Jordan, of Gad and of Ruben. On the extreme upper right of the map the word 'Terra' appears alone; one suspects that 'Terra Edom' was meant and that the right hand end of the map was unfinished.

Wey's map (Pls 5 and 6) clearly depends on Sanuto's, but the presentation and purpose of the two maps may have been very different. Sanuto's map was part of a treatise written to persuade Christian rulers to a new Crusade and seems intended to be realistic; Wey's map wears a different face:

> Here we are in a world of tradition and fable; as is usual in medieval maps the east lies at the top, but little attention is paid to proportion or direction; there is nothing precise or succinct in this cartography, but an abundance of decoration — towers, domes, turreted cities, out of whose windows look disproportionately large faces, some comely, some villainous. An elaborate and fanciful representation of a walled town stands for Jerusalem. The vivid blue with which the Dead Sea is coloured does not conceal the outlines of Sodom, Gomorrah, and the other cities of the Plain, prone but otherwise undamaged beneath the water. (Prestcott 1954, 75)

The first thing one observes about Wey's map is that it is a beautiful piece of artistic work. It presents a Holy Land of green (land) and blue (rivers) and red (cities, castles). Cisjordan and Transjordan appear united, not divided, by the river system. The contents of Wey's map are biblically based, and the world of the Crusaders is less evident than in earlier maps — for example, Mons Regalis is not shown. The text of the itinerary, however, is seriously practical, and includes much sound advice for pilgrim, including a glossary of useful vocabulary and phrases. Wey's map was surely also intended to be useful, locating places accurately enough by 15th century standards, and would have given the pilgrim or those who wished to follow the pilgrim's account at home a reasonable picture of the landscape, for all its errors, which are basically those of Burchard and Vesconte-Sanuto.

Gabriele Capodilista and his cousin Antonio Capodilista travelled with William Wey on the same pilgrimage in 1458; the whole party made up 197 people. Gabriele's map (Nebenzahl 1986, 53–56, pl. 18) is drawn in a much less exuberant, more restrained style than Wey's. It contains many of the same names; of its 41 Transjordanian place-names, it shares 32 with Wey's 51. Capodilista drew heavily on the Vesconte-Sanuto map, especially north of the Dead Sea, and on Burchard and perhaps Theoderich for the southern region. Capodilista, however, adds a completely new panel to the south, showing Lower Egypt and the Sinai region, complete with Mons Mousi (Jebel Mousa), the Fontes Mousi, Mons Sanctae Catherinae, and the Mons Vituli (where the Israelites worshipped the Golden Calf). Again, Capodilista's map is concerned to show places of biblical interest, but with them important Christian shrines, like the tomb of St Catherine on Mt Sinai. Capodilista, like Wey, follows Vesconte-Sanuto in his presentation of the Jordan and other rivers, and with Burchard and Vesconte locates Corozin on the north-east shore of the Sea of Galilee.

The hand-drawn maps of Wey and Capodilista stand at the end of a tradition going back via Sanuto-Vesconte to Burchard and his predecessors. The main concern of these cartographers was with the biblical narratives and the interests of the pilgrims; interest in the military exploits and castles of the Crusaders was fading. At the same time, the printed book was beginning to displace the manuscript; with the Renaissance there was a new interest in classical writers (including Ptolemy and his Geography), and with the coming of the Reformation, there was a new and different emphasis on biblical scholarship.

TABLE 1 Transjordanian toponyms found on fourteenth-/fifteenth-century maps

Florence c. 1300	*Sanuto-Vesconte 1321*	*Wey 1458 & 1462*	*Capodilista 1458*
Achama		Adama	Adama
Anor Seyr			
Aque Maron	Aq. Maron		Aque Maron
Ar civitas que nuc dicitur Areopolis	Ar, Areopolis	Arnon civitas	Areopolis
Baalgad	Baalgad	Baalgad	
Barcaranaea			
	Basan	Basan	
Bethin			

Florence c. 1300	*Sanuto-Vesconte 1321*	*Wey 1458 & 1462*	*Capodilista 1458*
Bethrecha quae nunc Betron dicitur	Bosra nunc Ydumea regio Bostron dicitur	Bosra quae nunc Bosera dicitur	Bostraham
Betsemut	Betsemut		
Campestria Moab (x 2)		Campestria Moab	
Cedar civitas	Cedar	Cedar civitas	Cedar
Convallis illustris		Convallis illustris	
		Coroniam	
Corozain	Corocaym	Corozaim	Corozaim
Damascus	Damascus	Damascus	Damascus
Caesarea Philippi/Belenas	Cesarea Phylippi/Belinas	Belinas civitas	Belyna
Dan unus de fontibus Iordanis	Dan	Dan fons unus de fontibus Iordanis	Dan
Decapolis regio			
	Dimidia Tribus Manasse	Dimidia tribus Manasse	
Effren	Efron	Effrem	Effren
	Eleale	Eleale	
Ermon	Hermon		Hermon
Esebon	Esebon		Eseb
Esumaim	Estonaym (?Betonaym)	Coronaim	
	Fagor	Phogor	M. Fagor
			Fontes Mousi
Gadera	Gadara	Godera	
Gomorra		Gomorra	Gomorra
Gorsa	Gerasa	Goraca	
Hamath			
Iabes	Iabes		
Iacor	Iacer	Jazer	
Introitus terre Hamach	Icadyt' emat	Introitus Emach	
		Iordanis fluvius	F. Iordanus
[Ior] secundus fons Iordanis	Ior	Ior	Ior
Manaym/Macheronta/ Aulon	Manay vel machero[n]ta n[un]c haylon	Manaim	Macheronta
Mare Galilee sive Gensaret vel Tiberiadis	Mare Tybiadis sive Galilee	Mare Galilee sive Tibiadis sive Cenereth sive Genezareth	Mare Galilee sive Tyberiadis
			M[are] Rubrum
Mons abarim	Abari[m]	Abarim	M[ons] Abarī
			M[ons] Falga
Mons Galaad	Galaad	Mons Galaad	M[ons] Galahad
Mons Hermon	Mos H'mon	Mons Hermon	Mōs Hermō
Mons Libanus		Mons Libani	Mons Libanus
			Mons Mousi
	Seyr [in Moab]		
Mons Seyr (in Tribus Gad nr Cedar)			
Mons Seyr (in Tribus Gad nr Ramath)	Iayr vel avot Seyr	Mons Seyr	
Mons Seyr (in Terra Edom)	Mos Seyr		
			Mons S[anctae] Caterine
			Mons Sinay
Mons Suanir [nr Mt Hermon]	Mons Sanyr {Seyr} coniungitur Damasco	Mons Sanyr	M[ons] Sanir

Florence c. 1300	*Sanuto-Vesconte 1321*	*Wey 1458 & 1462*	*Capodilista 1458*
			Mons Vituli
	Nebo	Nebo	M[ons] Nabo
Nemrin			
Pella	Pela	Pella	
Petra deserta	Pe[tra] deserta. . .dicitur Crac et mons regalis	Petra deserta	Petra deserti
Phanuel	Fanuel		
Phasgan	Fasgo		
Phiala	Fiala	Phiala	Fhpiala
		Ramoth Galaad	
Regio Traconitidis		Regio Traconitidis	
Secundus fons Jordanis		Fons secundus de fontibus Iordanis et dicitur Ior	
Segor		Segor	Segor
Sep[ulchru]m Iob	Sepulchrum Iob	Sepulchrum beati Job	S[epulchrum] Iob
		Sethir	
Socoth		Sochoe	
Sodoma		Sodoma	Sodoma
Suba, haec civitas dicitur turris Libani		Suba hec in Canticis dicitur turris Libani	
Sueta	Sueta		
Syria Damaschi		Syria Damasci	
Terra Edom	T[er]ra Edom	Terra [Edom]	
Terra Hamach			
		Terra Hus	
Terra quae dicitur Aman	T[er]ra a[mmo]ni[t]a[rum] n[un]c Aman	Terra Amon quae nunc Aman dicitur	
Terra sive regnum Saon Regis Esebon	Regia Seon Regis Esebon		
Torrens Arnon	Torrens Arnon	Torrens Arnon	Torrens Arnon
			Torrens Saul
Tribus Gad	Trib. Gad	Trib.Gad	
Tribus Neptalim	Trib. Neptalin	Trib. Neptalim	
Tribus Ruben	Trib. Ruben	Trib.Ruben	
Ydumea regio	Ydumea regio	Ydumea regio	
	Vallis salinarum/lacus asfaltidis/mare sodomorum vel maleficiorum vel mortuorum	Mare mortuum, mare sodomorum, lacus asbalti, vallis salinarum	Mare Mortū
		Zeboim	Seboi

CHAPTER THREE

THE FIRST PRINTED MAPS

LUCAS BRANDIS OF LUBECK

On 5 August 1475 Lucas Brandis in Lubeck published *Rudimentum novitiorum sive chronicarum historiarum epitome*, a fifteenth-century history of the world (for a study of the author and his sources, which include Burchard, see Schwarz 1888). Volume 1 contains the earliest known printed modern map of Palestine, with east at the top, and with individual towns or areas each presented as a separate hill or mountain. The map (Pl. 7) is a woodcut, with lettering printed from type inserts (Bagrow 1950, fig. 19; Nebenzahl, 60–61; Campbell, 1987, 146 and Fig. 61; see Röhricht, 1898, Tafel 9 for a version with type-set place-names but without geographical outlines; later versions of this map were published in Paris in 1488 and Lyons in 1491; see Campbell, 1987, 147–50). Transjordan is clearly divided from Cisjordan by two large stretches of water, the Sea of Galilee and the Dead Sea, connected by a very short River Jordan between them, in which Christ appears being baptised by John; Jor and Dan, the sources of the Jordan, are seen as one hill at the north end of the Sea of Galilee. Nebenzahl describes this map as Brandis' own woodcut version of Burchard's map (Nebenzahl, 62), and Ariel Tishby as 'a highly imprecise interpretation of Brandis de Schass of Burchard's text' (Tishby, 2001, 78); the main evidence for this connection seems to be that this map, like Burchard, sets Corosaim (Burchard's Chorazin) east of the Sea of Galilee. However, the list of Transjordanian names on Brandis' map does not closely correspond with those drawn from Burchard, and Burchard's error about Chorazin was repeated by Sanuto-Vesconte, Wey, Capodilista, and Breitenbach. Brandis was following what was by his day a well-established error. What is curious about this map is that in addition to Corosaim it locates to the east of the Sea of Galilee a number of other well-known Galilaean and Cisjordanian places — Tripolis, Cesarea Philippi, Caparnaum, Cana, Mons Tabor, Nazareth, Senereth, Naym, Betulia, and Cison torrens, Salim and Dothaim from further south.

In 1478, Brandis of Lubeck published two maps, one of the region of Jerusalem, and the other of Palestine, oriented with east at the top, and Antioch in the north, Egypt and the Red Sea in the south, in the *Prologus Arminensis in mappam Terrae Sanctae* (cf. Neumann 1885 [facsimile]; Röhricht 1898, Tafel 11; Campbell 1987, 121, no. 94, and fig. 60). These 'maps' (see Fig. 8) appear without geographical outlines, but with the place-names and certain *legenda* typeset in the appropriate place, like the version of the 1475 map published by Röhricht (1898, Tafel 9). Though Brandis here followed the style of his earlier map, and included many names from it, he made significant changes. The two sources of the Jordan, Ior and Dan, have become one, Iordanis oritur. Many places have been added, especially to northern Transjordan: Theman civitas, Phiala fons, Traconitidis regio, Naason, Neptalim, Mons Galaad, Sirie, Mesopotamia, Basan terra, Casbon, Desertum Bosor, Mons Hermon minor, Macheronta, Civitas Bosor, Effrem civitas, Terra Arnon. Tiberias has been transferred from the western to the eastern shore of the Sea of Galilee, but Tripolis, Cesarea Philippi, Capharnaum, Fons Ortorum, Cana, Nazareth, Mons Tabor, Chenereth, Cison torrens have reappeared west of the Sea of Galilee — though Corrosaim, Dothaim, Bethulia, and Naym remain stubbornly east of it. In the south Transjordan, Petra Deserti has been explicitly identified (cf. Burchard earlier) with Mons Reab [= Regalis].

ERHARD REUWICH AND BERNHARD VON BREITENBACH

Certainly the most dramatic fifteenth-century map known to us is that prepared by Erhard Reuwich, a

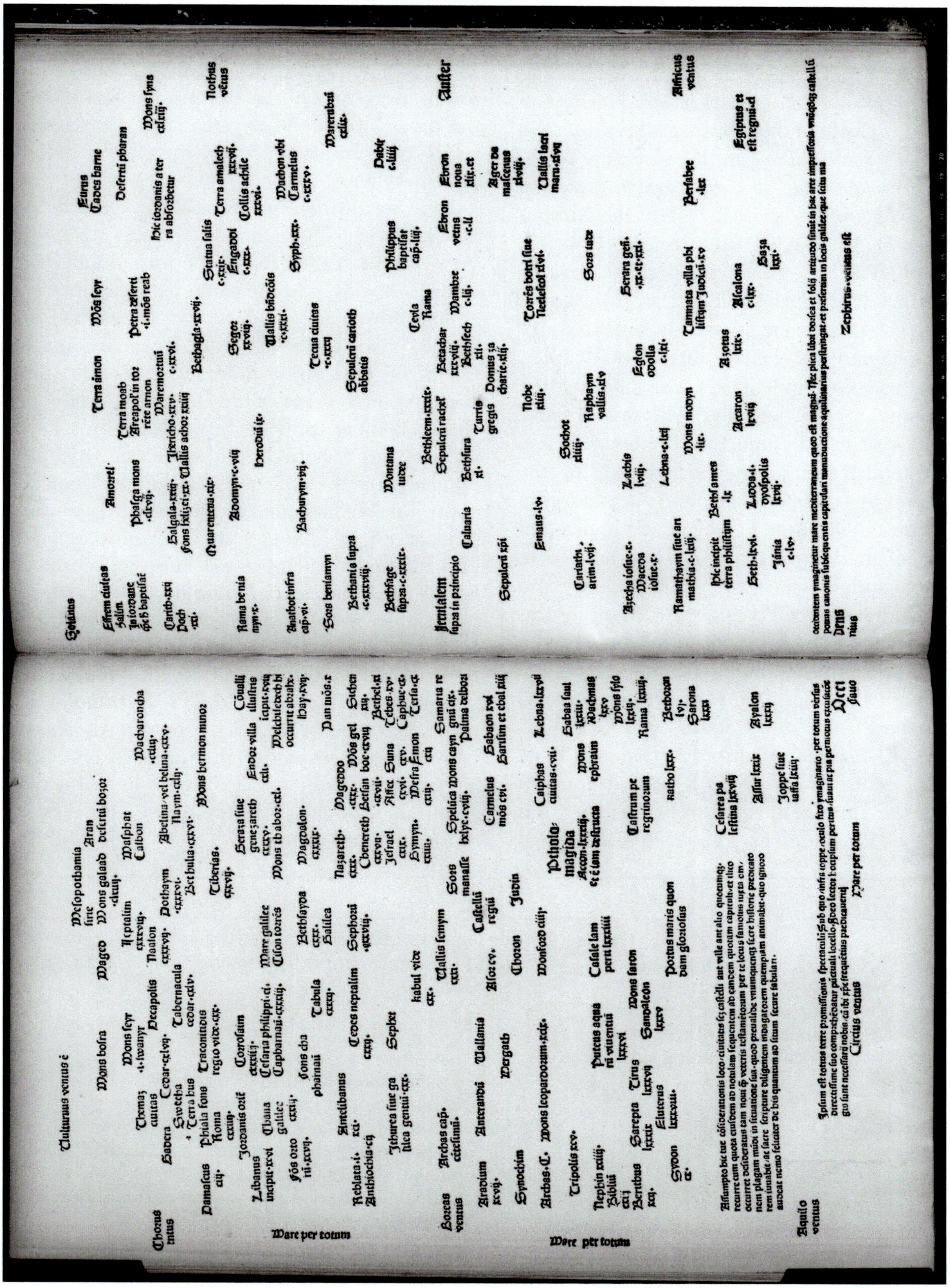

FIGURE 8. Typeset map from *Prologus Arminensis in Mappam Terrae Sanctae* (Lübeck, 1478) (facsimile in Neumann 1875). (*The British Library, London*)

professional artist from Utrecht, who accompanied Bernhard von Breitenbach, a canon of Mainz, on his pilgrimage in 1483–84. Breitenbach published an account of his journey, *Peregrinatio in terram sanctam*, in Mainz in 1486; the full title is *Sanctarum peregrinationum in montem Syon ad venerandum Christi sepulchrum in Jerusalem, atque in montem Synai ad divam virginem et matirem* [martyrem] *Katherinam opusculum.* (For the various editions, see H. W. Davies, 1911; see also Röhricht 1901, 129–35, Tafeln 1–3.) This is perhaps the first illustrated, printed travel book of the Holy Land. This particular pilgrimage is well documented, for the party included also Felix Fabri, Paul Walther, and Georg von Grumppenberg, who all left their own accounts of the journey. The main party confined its visit to Palestine, but Breitenbach and Reuwich extended their journey to include Sinai, the Red Sea, and the cities of Cairo and Alexandria in Egypt. Reuwich illustrated Breitenbach's account from sketches made on the journey reproduced in woodcuts (including views of Venice, Corfu and Rhodes; for a full list, see Davies, xxi–xxvii), the largest of which (27 × 127 cm; 10.5 × 50 in.) presented a panorama of the entire Holy Land, viewed from the west, looking east, from Damascus in the north to Alexandria and Cairo in the south, with CIVITAS IHERUSALEM displayed disproportionately large in the very centre, but reversed, with its eastern side facing the viewer. Imaginatively, Reuwich has presented the Holy Land (Fig. 9) as seen from a great height above the Mediterranean Sea in the west; the view includes Mons Libanus in the north, and Mons Synai and even Mecca in Arabia away to the south; away to the east beyond the Sea of Galilee, the River Jordan and the Dead Sea, can be seen the region of Transjordan, with a long line of mountains on the skyline.

Reuwich probably drew on the maps of his predecessors, especially Sanudo-Vesconte and William Wey. Reuwich's version of the Jordan and its tributaries is not exactly the same as theirs, but similar enough (the Sea of Galilee and the Mediterranean remain linked by river, but the River Arnon [Fluvius Oronaim] now rather more correctly flows into the north-eastern corner of the Dead Sea). To his credit, Reuwich drew the Red Sea with its two northern prongs, the eastern arm disappearing behind Mt Sinai and reappearing to its left, as the distant viewer above the Mediterranean would see it given suitably sharp eyesight. Though the Transjordan on Reuwich's map was compressed into the foreshortened far distance, the cartographer nevertheless managed to include a large number of the usual names. Corrozaym remains on the north-east shore of the Sea of Galilee. A comparison with the place-names presented by Burchard, Sanuto-Vesconte and Wey shows that most of Reuwich's Transjordanian place-names appeared earlier in Burchard and Sanuto-Vesconte, and almost all of them in William Wey. Reuwich, however, locates more Old Testament events in Transjordan:

Wey	*Reuwich*
	Hic fugit David ab facie filii sui Absalon
Effrem . . . conflictus inter exercitum David et Absalon	Hic fuit commissum proelium inter David et Absalon
Hic in campestribus Moab ultra Iordanem contra Jericho longo tempore steterunt filii Israel hic fuit editus [. . .]et Moyses mortuus et sepultus sub monte Abarim	Mons Abarim in quo sepultus dicitur Moyes
	Hic Aaron fuit mortuus et sepultus
	Via per quam filii Israel venerunt de monte Synay . . .
	Ubi Moyses minabat gregem ad interiora deserti
	Locus ubi datus fuit decalogus
	Via per quam filii Israel sicco pede transierunt mare rubrum
	Here David fled from the face of his son Absalon
Ephraim . . . battle between the army of David and army of Absalon	*Here was fought a battle between David and Absalon*

FIGURE 9. Section of E. Reuwich's map accompanying B. Breitenbach's *Sanctarum peregrinationum . . . opusculum*, 1486 (Davies 1911, pl. 27). (*The Board of Trinity College, Dublin*)

Wey	*Reuwich*
Here in the plains of Moab beyond Jordan opposite Jericho the children of Israel stayed for a long time. Here was proclaimed [the law?] And Moses died and was buried below the mountain of Abarim.	*Mount Abarim in which Moses is said to be buried.*
	Here Aaron died and was buried.
	The road by which the sons of Israel came from Mount Sinai
	Where Moses drove the flock to the middle of the desert
	The place where the Decalogue was given
	The way by which the sons of Israel crossed the Red Sea with dry feet.

Reuwich seems to have more interest in the location of biblical events beyond the Jordan than Wey. On the other hand, Reuwich is not particularly concerned with the events of the Crusading period; of Burchard's three Petras, Reuwich shows only Petra Deserti (at the site of Mons Regalis, though this name is not given). However, Reuwich is well aware of contemporary Christian shrines, marking the tomb of St Katherine at Mt Sinai, the return route of Christian pilgrims from Sinai to Cairo, and the monasteries of St Paul the first hermit and of St Anthony. He is also aware of contemporary religio-political realities, marking on the Red Sea the port for the Indian trade and further down the Sea the tomb of Mahomet at Mecca in Arabia.

This map, for all its dramatic presentation, marks no great advance in topographical knowledge of the Transjordan, being completely dependent on its predecessors. Clearly Reuwich is interested in the greater geographical picture — the very presentation of his map is a testimony to that — but his evidence for the Transjordan is more biblical and literary than genuinely topographical. First-hand knowledge, however, is clearly visible in Reuwich's presentation of the southern end of his map — the region of Sinai, the Red Sea, and the Nile Delta, with Alexandria and Cairo. Reuwich's picture of Mount Sinai suggests at once that he has seen Jebel Musa, and his location of it in the Sinai triangle between the arms of the Red Sea and the Gulf of Aqaba shows that he understands the general geography. This map must be recognised as a remarkable achievement.

PTOLEMY

When Breitenbach and Reuwich published the *Peregrinatio*, with its panoramic map, in 1486, the Ptolemaic maps of Syria-Palestine were already in print. The first printed edition, in Latin, appeared at Bologna in 1477; it was followed by another Latin edition printed in Rome in 1478 (a later edition from the same plates appeared in 1490). In 1482, an edition in Italian edited by Berlinghieri was printed at Florence (cf. the facsimile published in *Theatrum Orbis Terrarum* III.4, Amsterdam 1966), and an edition in Latin edited by Nicolaus Germanus was printed at Ulm (a later edition from the same plates appeared in 1486) (for a full list of early printed editions, cf. Stevens 1908; Moreland and Bannister 1989, 301–02). Breitenbach and Reuwich may have known the Ptolemaic maps, either through manuscript or printed versions, though their map seems to belong to the tradition of Burchard, Sanudo-Vesconte and Wey. However, as is well known, Palestine appears in the Ptolemaic corpus on two very different maps. The first map is the *Quarta Asiae tabula* (Pl. 10); this map belongs to the original collection of 26 Ptolemaic regional maps, and covers the whole of Syria, Palestine, Arabia and Mesopotamia. This map appeared in the Bologna edition (1477), the Rome editions (1478 and 1490), the Berlinghieri edition from Florence (1482), and the Ulm editions (1482, 1486). The second map on which Palestine appears is the *Palestina moderna et terra sancta* of Berlinghieri's *Geographia* from Florence (1482), which appears as *Tabula moderna terre sancte* in the 1482 and 1486 editions from Ulm. (It appears also under the same title in the copy of the 1490 Rome edition held in the Library of Trinity College, Dublin [Pls 8 and 9], though its presence there is not noted by Campbell [1987] or Moreland and Bannister [1989].) Lastly, a version of it is attributed to Rosselli in Florence *c.* 1492 (cf. Campbell 1987, 70 and fig. 66; Almagia 1951, 27–34).

These two maps, the *Quarta Asiae Tabula* and the *Tabula Moderna Terrae Sanctae* are very different. The *Quarta Asiae Tabula* is clearly drawn from the Graeco-Roman tradition, naming provinces and cities of the Roman east, and the second, as its very title suggests, from the Christian tradition, naming places known from the Bible and from the experience of Crusaders and pilgrims. Thus, if we examine the city-names of the provinces of Curva (or Cava) Syria and Arabia Petraea in the *Quarta Asiae Tabula*, we find that out of about 50 names, only Gerasa, Gadara (cf. Mk 5.1, Mt. 8.28), Esbuta (Hesbān) and Characmoba connect with the biblical tradition. The rest are known to us from Josephus, Pliny, the Antonine Itinerary and the Peutinger Table, with the exception of a group of names of places in Arabia Petraea peculiar to Ptolemy (Adra, Mesada, Sunatha, Magura, Medina, Lidia, Rabimathon, Cletaro, Thano, Calguia, Pharna villa, Munychiatis regio).

If we examine the names of the Tranjordanian region in the *Tabula Moderna* maps, we find that the names are drawn almost entirely from the biblical tradition. This is not surprising, since the Berlinghieri map 'follows the fourteenth-century model of that accompanying manuscripts of Marino Sanudo's *Liber Secretorum Fidelium Crucis*' (Skelton 1966, xii). Thus in the *Tabula Moderna* maps the Transjordanian region is divided between Trachonitis (cf. Luke 3.1), the Israelite tribes of Naphtali, half Manasseh, Reuben and Gad, and the land of Edom. Bosra of Edom has apparently been confused with Bosra in Syria, Idumaea being located near Damascus. The fountain of Phiale has appeared from Josephus (*BJ* 3.510). Engedi, strangely, appears east of the Dead Sea. Brief references appear to biblical stories set east of the Jordan, e.g., to David and Absalom, to Balaam's cursing of the people of Israel, and to Moses' delivering the book Deuteronomy, to the kingdoms of Sihon and Og, and to Job's association with the land of Uz. There is a brief reminder of the Crusades in the location of Mons Regalis (Montreal = Shobek, identified with Petra deserti; cf. Burchard). Reference to the Roman division of Arabia appears in an abbreviated reference to Arabia S[e]c[un]da. Clearly the purpose of the *Tabula Moderna Terrae Sanctae* maps was to meet the needs of biblical readers and pilgrims to the Holy Land, rather than to present contemporary geography and topography. Ptolemy's *Quarta Asiae Tabula* map, however, was not concerned with the Bible. It is true that there is reference to Sinai mons or Mons Sina on the versions of Codex VF from Naples and Codex Urbanas Graecus 82, but there is no reference to Mt Sina(i) in the *Nomenclator Ptolemaicus* (in Ortelius' *Theatrum Orbis Terrarum*, 1579) or in the early printed versions of Ptolemy.

LUCAS CRANACH THE ELDER

The *Map of the Holy Land* by Lucas Cranach the Elder, recently dated by A. Kunz to a date between 1522 and 1525 (Kunz 1995, 123–44), follows in general outline the *Quarta Asiae Tabula* of Ptolemy, perhaps using extra information from the *Palestina Moderna* map (Kunz 1995, 131). For the Transjordanian region proper Cranach's map (Fig. 10) offers little compared with previous maps. From Hermon in the north, moving southwards Cranach names Dan, Cesarea Philip[i], [Lake] Genesara, Gilead, Esbon, [the river] Iaboc (strangely placed south of Esbon), Blachfeld Moab, [Mount] Pasga, [the river] Arnon, [the river] Sared (both flowing apparently into the mountains on the east shore of the Dead Sea). The tribal lands of Manasse, Gad and Reuben are defined by boundaries, and the lands of the Ismaelites, Midianites, Moabites and Ammonites indicated. The main feature of Cranach's map, however, taking up the whole of the lower left foreground, is the route of the Israelites' march through the wilderness marked 'Seir' and 'Edom'. Cranach constructs this map not from Numbers 33, as later mapmakers do, but mostly from the story as recounted in the books of Exodus and Numbers, illustrating the more important events with small vignettes (see Kunz 1995, 132–34). He begins with Sochot (Succoth) (Exod. 12.37), shows the Israelites crossing the Sea pursued by the Egyptian chariotry, and then locates Ethan (Etham) (Exod. 13.20), Marath (Marah) (Exod. 15.23), Elim (Exod. 15.27), Sur (wilderness of Shur) (Exod. 15.22), Sin (wilderness of Sin) (Exod. 16.1), with an illustration of the manna from heaven on the ground, Rephedim (Rephidim) (Exod. 17.1), with an illustration of Moses striking the rock, Sinai (Exod. 19.1), with an illustration of Moses on top, and the golden calf being worshipped below, Cades (Kadesh) Barnea and the wilderness of Pharan (Paran) (Num. 13.26), with an illustration of the result of Korah's rebellion (?), Zin (and Hebron) (Num. 13.21), (Mount) Hor (Num. 20.22), Salmona (Num. 33.41), Punon (Num. 33.42), with an illustration of the serpents from Num. 21.6–9, Oboth (Num. 21.10, 33.43), Abarim (Iye-abarim) (Num. 21.11, 33.44), (the river) Sared (Zered) (Num. 21.12), (the river) Arnon (Num. 21.13), (the river) Iaboc (Num. 21.24), and the Blachfeld Moab (plains of Moab) (Num. 22.1). Cranach also locates Mount Pisga (Num. 21.20) and the city of Ar (Num. 21.28). This reconstruction of the Israelite route through the wilderness, presented as a winding curve through the mountains of Edom, set a precedent that would be followed by biblical mapmakers through the sixteenth century. Cranach's concern is to illustrate the biblical story, to which he

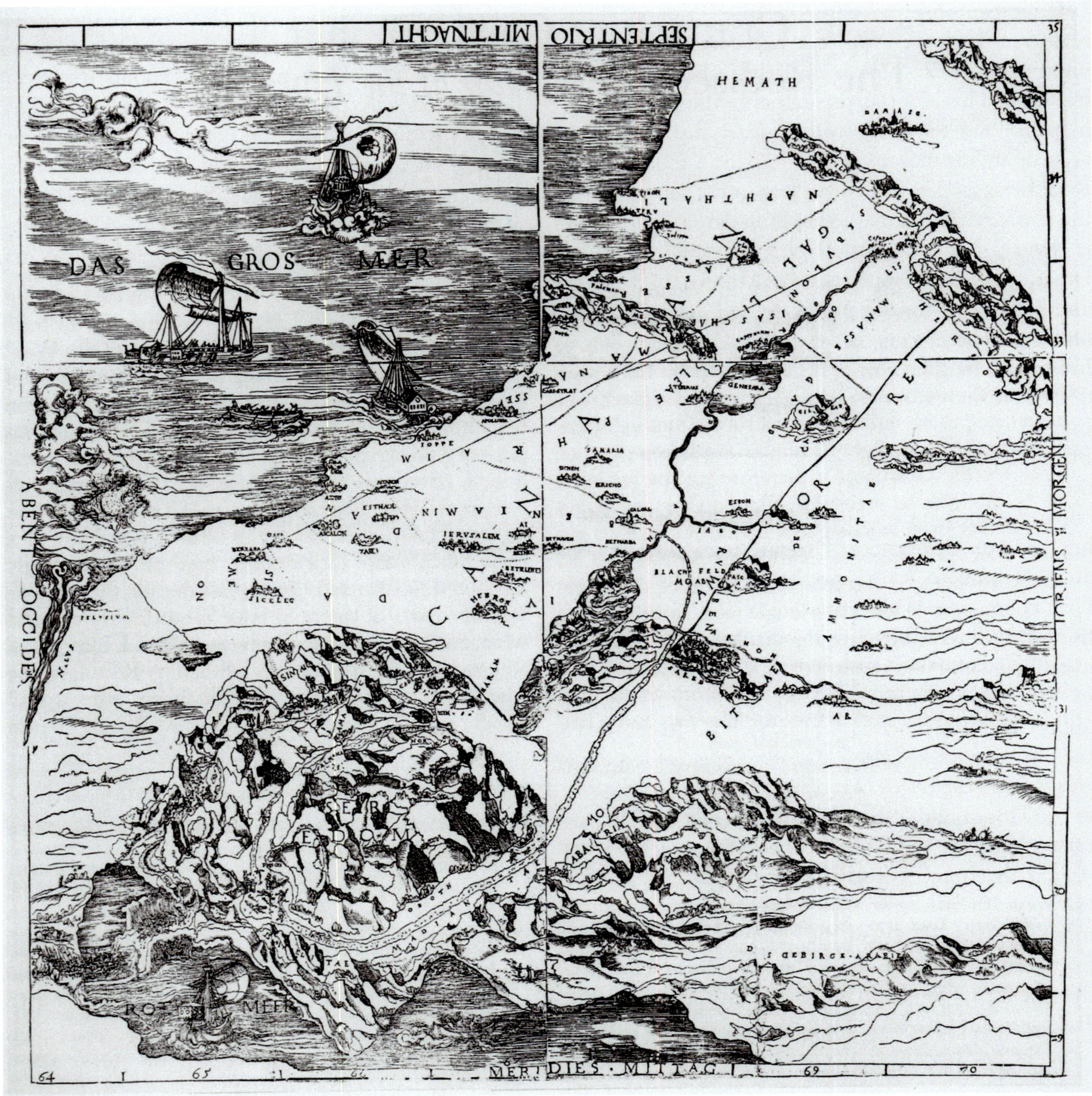

FIGURE 10. L. Cranach the Elder, Map of the Holy Land, *c.* 1522–25 (Kunz 1995). (*Armin Kunz and Print Quarterly 12.2 (1995), 124*)

has converted the Ptolemaic map. Cranach's map was used in 1525 in an edition of Luther's translation of the Old Testament published by Christoph Froschauer of Zurich, and in 1526 in a Dutch Bible published by J. Liesveldt in Antwerp (Delano-Smith 1987, 4–5).

These early printed maps differ in presentation, content and intention considerably. Brandis published his first, somewhat schematised map (1475) in a volume dedicated to world history, and for the details of the Holy Land and Transjordan he draws on fourteenth- and fifteenth-century sources, but without taking much care to locate places accurately; his later map (1478) seems to have made some attempt at correction. Reuwich presented an impressive landscape-view map that combined information from earlier maps with his own observation as a traveller. The several printings of the maps associated with Ptolemy's *Geography* offered two distinct maps, one drawn from the original Ptolemaic tradition of the Roman empire in the second century, and the other from the tradition of late medieval biblical scholarship. Lastly, Cranach's map uses the earlier Ptolemaic map to present the biblical account of the Exodus and wilderness wanderings. Apart from the input of the Ptolemaic tradition, these early printed maps reflect the biblical interests — and in Cranach's case, the Reformation interests — of their time. They also demonstrate clearly the potential for the cartographer of the newly developed skills of the printer.

TABLE 2. Transjordanian toponyms on maps from 1475–1525

Rudimentum Novitiorum 1475	*Prologus Arminensis 1478*	*Reuwich & Breitenbach 1486*	*Ptolemy Quarta Asiae Tabula Rome 1490)**	*Ptolemy Tabula Moderna (Rome 1490)*	*Lukas Cranach 1522–1525(?)*
		Mons Abarim			Abarim
		Adama			
Abelina vel Belina	Abelina				
	Amonite				Ammonitae
Amorrei	Amorrei				Amorrei
				Aque Moron	
					Ar
				Arabia secunda	
Aran					
Areopolis	Areopoli	Areopolis castrum			
Arnon				Arnon	Arnon
		Avota			
		Baalag			
				Balgat	
Basan terra				Basan	
Bethulia	Betulia				
					Blachfeld
				Bosra	
				Bosra nunc	
				Idumea	
				Regio Bosra	
	Cades Barne				
				Campestria Moab	
	Cana				
	Caparnaum				
Casbon					
Cedar, Tabernacula	Cedar et tab. eius	Cedar		Tabernacula	
Cedar				Cedar	
	Cesaria Philippi		Cesarea	Cesare	Cesaria Philip[pi]
				Philippi	
Civitas Bosor				Besor civitas	
				fugitivorum	
Corrosaim	Corosaim	Corrozaym		Corosaim	
Damascus	Damascus	Damascus	Damascus	Damascus	Damascus
	Dan	Dan	Dan	Dan	Dan
Decapolis	Decapolis				
	Deserta Moab				
	Deserta Suetha				

Rudimentum Novitiorum 1475	*Prologus Arminensis 1478*	*Reuwich & Breitenbach 1486*	*Ptolemy Quarta Asiae Tabula Rome 1490)**	*Ptolemy Tabula Moderna (Rome 1490)*	*Lukas Cranach 1522–1525(?)*
Desertum Bosor					
Desertum Pharam	Desertum Phares				Pharam
				Diadada	
				Dibon	
Dothaym	Dothaim				
				Terra Edom	Edom/Seir
Effreym civitas		Effreym		Effrem	
				Eleale	
	Endor				
				Engedi [on E shore of Dead Sea]	
				Esbon (twice)	Esbon
				Fanuel	
		Fluvius Oronaym		Coronaim	
	Fons Ortorum				
		Tribus Gad		Tribus Gad	Gad
Gadera		Gadera	Gadara		
		Gebin			
		Gera	Gerasa	Gerhasa	
			Lacus Genesare		Genesara
				Getasame	
				Golum vel Gauleon	
		Gomorra			
		Iabes		Iabes	
				Iaboc	Iaboc
		Ydumaea regio			
		Introitus Emath		Introitus in amath	
	Ior		Ior		
Iordanis		Iordanis fluvius	Iordanis flu.	Iordanis flu.	Iordanis
					Ismaelitae
		Jaser		Iaser	
		Jerasa			
Macheronta				Mahanaim vel	
				Macheronta nunc ayon	
					Madiamitae
Maged					
				Manasses	Manasse
Mare Galilee		Mare Galilee vel	Lacus Tiberiadis	Mare Galilee	
		Mare Tiberiadis		Mare Tiberiadis	
	Mare mortuum	Mare mortuum	Mare mortuum	Mare mortuum	
				sive Sodomorum	
Masphat				Masfat	
	Melchisedech				
					Mobitae
Mons Bosra	Mons Bosra				
Mons Galaad					
Mons Hermon	Heremo	Hermon		Hermon mons	Hermon
minor				& Hermon (minor)	
	Mons leopardorum				
		Mons Libanus	Libanus mons	Mons Libanus	
		Mons Nebo			

Rudimentum Novitiorum 1475	*Prologus Arminensis 1478*	*Reuwich & Breitenbach 1486*	*Ptolemy Quarta Asiae Tabula Rome 1490)**	*Ptolemy Tabula Moderna (Rome 1490)*	*Lukas Cranach 1522–1525(?)*
		Mons Oreb			
		Mons Phagor			
Mons Seyr vel Swanyr	Mons Seyr			Mons Seyr	
	Mons Syn	Mons Synay			
	Mons Tabor				
Naason					
Naym	Naym				
	Nazareth				
Neptalim		Tribus Neptaim		Tribus Naptalim	
				Norim	
					Pasga
		Pella	Pella	Pella	
Petra deserti id est Mons Reab	Petra deserti	Petra deserta castrum	Petra	Petra deserti vel mons regalis	
Phiala fons		Fons Phiala		Phiala fons	
		Ramath Galaath		Ramat Galadanus	
		Tribus Ruben		Tribus Ruben	Ruben
Salim	Salim				
				Sabania	
					Sared
	Senereth				
		Sepulchrum Iob		Sepulchrum Iob	
		Sepulchrum Sanctae Catherinae			
		Seboim			
		Segor			
		Set [=Shittim]		Serim	
				Siba	
Sirie					
		Sochot		Sochot	
		Sodoma			
Swecha		Sueta		Sueta	
Terra Hus	Terra Hus	Terra Hus		Terra Hus	
Theman civitas					
	Torrens Cison				
Traconitidis regio		Traconitidis regio		Regio Traconitidis	
	Tripolis				
				Zair vel hayot	
					+ mansiones

* For this column only those toponyms are given which correspond to other toponyms in the table. For the full list of toponyms given in Ptolemy's *Quarta Asiae Tabula*, see the separate table. The Ptolemaic maps used for the purposes of the chart above are from the Roman edition of 1490 in the Library of Trinity College, Dublin.

TABLE 3. Transjordanian toponyms from the early editions of *Quarta Asiae Tabula*

Codex VF Naples pre-1470	*MS N.Germano Florence 1474 Nebenzahl, 16–17*	*Bologna 1477 (ed. Lynam 1941)*	*Berlinghieri Florence 1482*	*Ulm 1482, 1486 (BL IC.9303)*	*Rome 1490 (TCD)*	*Rome 1490 facsimile Nordenskjold*	*Codex Urbinas Graecus 82 Leiden 1932, Ed. J. Fische (transliterated)*	*Nomenclator Ptolemaicus Antwerp 1569 (TCD PP.a.34)*
Damascus	Damascus	Damascus	Damasco	Damascus	Damascus	Damascus	Damaskos	Damascus
antilibanus	Antilibanus mons	Antilibanus m.	Antilibanus mo.	Antilibanus mons	Antilibanus mons	Antilibanus mons	Antilibanus oros	Antilibanus
Jordan flu.	Ior, Dan Iordan flu	Iordanis F.	Iordano fl.	Iordanis fl.	Dan, Ior f. Iordanis fl.	Jordanis fluvius	Iordanis pot[amos]	Iordanis flu.
Genesar lacus	Lacus genesar		genesarete lago	Lacus Genesar	Lacus genesare	Lacus genesar		Genesaritis lacus
Taiberias lacus	Lacus tiberiadis		tiberiadi lago	Lacus Tiberiadis	Lacus tiberiadis	Lacus tiberias	Tiberias limne	Tiberias
Aspaltidis lacus	Lacus aspaltis		asphaltidi lago	Lacus Asphaltis	Lacus aspaltis	Aspaltidis lacus	asphaltis limne	
Curva siria	Curva Syria	curva syria	curva syria	Curva Syria	Curva Syria	Curva Syria	Koile Syria	Syria cava
Capitolia	capitolia	capitoli	capitolia	capitolia	capitolia	capitolias	capitolias	Capitolias
	abida	abida	abida				abida	Abida
Hippos	hippus	ippus ipus m.	hippo mō	hippus	hippus	Hippos	hippos	Hippos
Hippos m.						Hippos mos		Hippos mons
gadara	gadara	gadara	gadara	gadara	gadara	gadara	gadara	Gadora
adra	adra	adra	adra	adra	adra	adra	adra	Adra
diu	dium	dion	dio	dium	dium	dium	dion	Dium
pella	pella	pella	pella	Pella	pella	pella	pella	Pella
gadora	gadara	gadoria	gadora	godara	godara	gadora	gadora	Idara
filadelfia	philadelfia	philadelfia	philadelphia	philadelfia	philadelfia	philadelphia	philadelphia	Philadelphia
scitopolis	scytopolis	sitopoli	scythopoli	scitpolis	scitopolis	scythopolis	skythopolis	Scythopolis
gerassa	gerasa	gerassa	gerasa	gerasa	gerasa	—	gerasa	Gerasa
canata	canata	canata	canatha	canata	canata	—	kanatha	Canatha
In Palestina								
Iudea								
cosmos	cosmos	cosmos	cosmo	cosmos	cosmos	cosmos	kosmos	Cosmus
libias	libias	libias	libia	libias	libias	libias	libias	Libias
caliroe	calliroe	caliroe	caliroe	callirhoe	calliroe	calliroe	kalliroe	Callirhoe
gazorus	gazorus		gazoro	gazorus	gazorus	gazorus	gazorus	Gazorus-
Petraea Arabia	Arabia petrea	Arabia Petrea	Arabia Petrea	Arabia Petrea	Arabia petrea	Arabia Petraea	arabia petraia	Arabia Petraea
epicaerus	epiceros		epicaro	epiceros	epiceros	epicaeros	epikairos	Epicaerus
adra	adra	adra	adra	adra	adra	adra	adra	Adra
Bostra legio	Bostra legio	bostra legio	Bostra legio	Bostre legio	Bostra legio	Bostra legio	bostra	Bostra Legio
mesada	mesada	mesata	mesada	messada	messada	mesada		Sebunta (?)
sunatha	sunatha	sunata	surattha	sunatha	sunatha	subattha	surattha	Surattha
anitha	anitha		anitha		—	anitha	anitha	Anitha
esbuta	esbuta	ebuta	esbuta	esbuta	esbuta	esbuta	esbuta	
Sisa	ziza	zisa	ziza	ziza	zisa	zyza	ziza	Ziza
magura	maguza		maguza	magusa	magusi	maguza	maguza	Maguza
medina	medina		madava	mediā	mediā	medava	madaua	Medava
lidia	lydia	lida	lydia	lydia	lydia	lydia	lydia	Audia
rabinathon	rhabinathon		radmathmo	rhabinato	rhabinato	rhabmathmom	rabmathmom	Rhabmathmoma
cletaro	cletaro	celetara	cletharo	cletaro	cletaro	cletharro	kantharro	Cletharro
thano		thana	thana			thana	thura	Thoana
calgina	chalgina	calgiua	calguia	chalguia	chalguia	calguia	kaligia	Calguia

Codex VF Naples pre-1470	*MS N.Germano Florence 1474 Nebenzahl, 16–17*	*Bologna 1477 (ed. Lynam 1941)*	*Berlinghieri Florence 1482*	*Ulm 1482, 1486 (BL IC.9303)*	*Rome 1490 (TCD)*	*Rome 1490 facsimile Nordenskjold*	*Codex Urbinas Graecus 82 Leiden 1932, Ed. J. Fische (transliterated)*	*Nomenclator Ptolemaicus Antwerp 1569 (TCD PP.a.34)*
maliata	maliatha	malita	maliattha	maliattha	maliattha	maliattha	maliatth	Maliatha
Soracene regio ®				Saraceni			sarakene chora	Saracena
Petra	Petra	petra	petra	petra	Petra	Petra	petra	Petra
soara	zoara	zoara	zoara	zoara	zoara	zoara	zoara	Zoara
necla	necla	necla	necla	necla	necla	necla	nekla	Necla
motha	moca	motha	moca	moca	moca	moca	moka	Moca
Pharonite ®	Pharanite ®		pharanite	phanarite	pharanite	PHARANITE	pharanita	Pharanitae
lisa	lysa	lisa	lysa	lysa	lysa	lysa	lysa	Lysa
rhuba			gubba			Gucca	gubba	Guba
zanatha	zanata	zanatha	zanatatha	zamatha	zamatha	zanata	zanaath	Zanaatha
gipsaria	gipsaria		gypsaria	gipsaria	gipsaria	gypsaria	gypsaria	Gysaria
gerassa	gerassa	gerasa	gerasa	gerassa	gerassa	gerasa	gorasa	Gerasa
Helena villa	Elena villa	helana	elana villa	Elana villa	Elena villa	Elana villa	elana	
Melane mons	Melanes vel nigri montes		melani et negri montes	Melani vel nigri montes	Melanes vel nigri montes	Melanes montes		Melani mons
Sinai mons								
Pharna villa	pharnevilla	farna v.	Pharna villa	Pharna villa	Pharne villa	Menphara villa		
			coraca				korake	coraca
promontorium		pm				Cathapharna promontorium		Promontorium Pharan
caracmoba	characmoba	caracmoba	characoba	carachmoba	carachmoba	carachmoba	charakmoba	Characoma
avara		avara	avara	avara	avara	avara	auara	Avara
adros	adros	adros	adro	adro	adro	adros	adrou	Adri
			ratheni	rateni	Reteni	RATHENI	rathenoi	Rhaetani regio
SINUS ARABICI PS		Pars sinus arabici et mare rubrum	arabico seno	Sinus arabicae pars	Sinus arabici pars	SINUS ARABICI PARS	Arabikou kolpou muchos	
			elanite seno	Sinus heleniticus	Sinus haleniticus	SINUS ELANITIS	Elanites kolpos	
eboda	eboda	eboda	eboda	eboda	eboda	eboda	ebodas	Eboda
Munichiatis regio ®			Munychiate convento			MUNYCHIATIS REGIO	Munichiatis chora	Munichiatis regio

TABLE 4. Transjordanian toponyms from the fifteenth-century *Tabula Moderna*

Ptolemy Tabula Moderna Terre Sancte *Ulm 1482, 1486*	*Berlinghieri* Palestina Moderna et Terra Sancta *Florence 1482*	*Bonifacio Florence* Palestine (untitled) *'not later than 1490'*	*Ptolemy* Quarta Asiae Tabula *Rome 1490**	*Ptolemy* Tabula Moderna *Rome 1490 (TCD Library)*	*Rosselli(?)* Palestine (untitled) *Florence 1492*
		Abitarum m.			
	Abarin mo.				Abraym mons
		Adara			
	Adama	Adama			
		Aenum			
	Terra Amon			Terra Amon	
Aque Moron	Aque maron	Aquae Moron		Aque Moron	Aqua Morom
	Arabia prima				
		Arabiae Petré pars	Arabia Petraea		
Arabia secunda				Arabia secunda	
	Arnon			Arnon	
		Aropolis			
Astaroth		Astaroth			
		Azer			
Balgat	Baalgad			Balgat	Balgat
		Balneas			
				Basan	Basan
		Bathanaea			
		Betiesimoth			
Besor civitas fugitivorum	Besor	Bosor [in Moab]		Besor civitas fugitivorum	Betsorcivitas fugivitorum
Bosra	Bosra			Bosra	
Bosra nunc Idumea	Idumea regione	Idumea		Bosra nunc Idumea	
Regio Bosra	Hora Borseron			Regio Bosra	
		Cades Barne			
				Campestria Moab	Campestria Moab
Cedar	Cedar	Cedar		Cedar	Cedar
Cesarea Philippi	Cesarea di Philippo		Cesarea	Cesare Philippi	Cesarea filippi/Belina
	Corosa				
Corosaim	Corasaim			Corosaim	Corosaym
Damascus	Damasco	Damascus	Damascus	Damascus	
Dan	Dan fõte	Dan	Dan	Dan	Fons Dan
		Dabir			
Decapolis	Decapoli				
				Diadada	Diadamba
Dibon	Dibon			Dibon	
Terra Edom		Edom		Terra Edom	
Effrem	Effrem	Ephrem		Effrem	Effré[m]
Eleale	Eleale			Eleale	Eleale
Engadi				Engedi [on E shore of Dead Sea]	Engadi
		Esdrelon			
Esebon	Esebon	Esebon		Esbon (twice)	Esebon
					Fagor
	Fagor monte				Fagor mons
Fanuel	Fanuel			Fanuel	Fanuel
	Falga mo.				Fasga mons
	Oronaim	Oronaim		Coronaim	Coronaym
Tribus Gad				Tribus Gad	Tribus Gaad

Ptolemy Tabula Moderna Terre Sancte *Ulm 1482, 1486*	*Berlinghieri* Palestina Moderna et Terra Sancta *Florence 1482*	*Bonifacio Florence* Palestine (untitled) *'not later than 1490'*	*Ptolemy* Quarta Asiae Tabula *Rome 1490**	*Ptolemy* Tabula Moderna *Rome 1490 (TCD Library)*	*Rosselli(?)* Palestine (untitled) *Florence 1492*
	Gadara	Gadara	Gadara		Gadara
	Galaad monte				Galat mons
Gelon vel Gaulon	Golon			Golum vel Gauleon	Galon
Gerhasa	Gerhasa	Gerasa	Gerasa	Gerhasa	Gerhasa
			Lacus Genesare		
Getasame	Getasame			Getasame	Gerhasume
	Gomorra	Gomorra			
Hermon (mi)	Hermon (mi)			Hermon (mi)	
Iabes		Iabes Galaad		Iabes	Iabes
		Nemus Iaboc		Iaboc	
Introitus in emath				Introitus in amath	
Ior		Lora	Ior	Ior	
Iordanis fluvius	Iordano fiume	Iordanis F.	Iordanis flu.	Iordanis flu.	Iordanis fluvius
Iaser	Iasem			Iaser	
		Madean			
Manaim vel Macherota, Ayon	Manaim vel Macherota	Macherunta		Mahanaim vel Macheronta nunc ayon	Macronta nunc ayon
Tribus Manasses ultra Iordanem	Tribo Manasse di lada Iordano			Tribus Manasses ultra Iordanem	Tribus Manases
Mare Galilea Mare Tiberiadis	Mare di Galilee	Mare Galilaea sive Tyberiadis	Lacus Tiberiadis	Mare Galilee Mare Tiberiadis	Mare Gallilaeae
Mare mortuum sive Sodomorum seu lacus aspalti	Mare morto	Mare mortuum	Mare mortuum	Mare mortuum sive Sodomorum	Vallis silvarum sive Sal maris. Lacus aspaltis sive mare Sodomorum vel maledictum seu mortuum
Masfat	Maspath	Masphoe		Masfat	Masfat
	Moab				
Hermon mons	Hermon mo.			Hermon mons	Hermon mons
Mons Lybanus	Libano mo.	Libanus montis	Libanus mons	Mons Libanus	Libano mons
	Suanir mo.				Senir mons
	Seyr mo.			Mons Seyr	Seyr mons
					Nabon
	Nabo m.				Nabon mons
Tribus Neptalim				Tribus Naptalim	
Norim				Norim	Norym
Pella	Pella	Pella	Pella	Pella	Pella
Petra	Petra deserta	Petra regalis	Petra	Petra deserti vel mons regalis	Petra
Fons Phiala	Phiala fōte			Phiala fons	Fons fiala
Ramoth	Ramoth Galaad	Ramot		Ramat Galadanus	Ramot
Tribus Ruben	Tribo Ruben			Tribus Ruben	Tribus Ruben
	Sabama			Sabania	Sabama
		Sandinella			
	Sarech torete				
		Save			
	Segoin	Seboim			
	Segor	Zoara			
	Seir				

Ptolemy Tabula Moderna Terre Sancte *Ulm 1482, 1486*	*Berlinghieri* Palestina Moderna et Terra Sancta *Florence 1482*	*Bonifacio Florence* Palestine (untitled) *'not later than 1490'*	*Ptolemy* Quarta Asiae Tabula *Rome 1490**	*Ptolemy* Tabula Moderna *Rome 1490 (TCD Library)*	*Rosselli(?)* Palestine (untitled) *Florence 1492*
Sepulchrum Iob	Sepulcro di Iob	Sepulcrum Iob		Sepulcrum Iob	Sepulcrum Iob
Serim	Serin			Serim	Serim
Siba	Siba	Siba		Siba	
Sochot	Sochor	Socot		Sochot	
	Sogdoma	Sodoma			
Sueta	Sueta	Sueta		Sueta	
		Syriae pars			
Tabernacula Cedar	Tabēna	Tabernacula Cedar		Tabernacula Cedar	Tabernacula Cedar
	Terra Hus			Terra Hus	Terra Hus
		Traconi			
	Traconitae regione	Traconitidis		Regio Traconitidis	Regio Traconitis
Turris Libani	Tore di Libano			Turris Libani	
Sayr vel Iavoth	Jair vel Iamoth			Zair vel hayot	oorit (??) Iamoth

* For this column only those toponyms are given which correspond to other toponyms in the table. For the full list of toponyms given in Ptolemy's *Quarta Asiae Tabula*, see the separate table. The Ptolemaic maps used for the purposes of the chart above are from the Roman edition of 1490 in the Library of Trinity College, Dublin.

CHAPTER FOUR

JACOB ZIEGLER: *QUAE INTUS CONTINENTUR* AND *TERRAE SANCTAE . . . DESCRIPTIO*

THE NEXT MAJOR SCHOLAR to produce maps of the Transjordan was Jacob Ziegler. Ziegler was born in Landau about 1470, and died in Passau in August 1549. In his somewhat restless life he lived in Moravia, Hungary, and Italy as well as Vienna, where he taught theology. According to the *Allgemeine deutsche Biographie* (Leipzig, 1875–1912), 'Er war eine Kampfnatur; litterarische, politische und religiose Streitigkeiten hatte er immer auszusechten', and he published five books against the Waldensians in 1512, and a 'Libellus adversus Jacobi Stunicae maledicentiam, pro Germania' in 1522. In the 1520s he appears to have allied himself to Erasmus of Rotterdam, and published anti-papal pamphlets. In 1531 Ziegler settled in Strassburg, apparently espousing the Protestant cause, but in 1533 left Strassburg and returned to the Roman fold, in which he remained until his death. In 1536 he published a work *De solidae sphaerae constructione*, and shortly before his death he published *In C. Plinii de naturali historia lib. secundum commenta[ria]. Quibus difficultates omnes Plinianae astronomicae praesertim explicantur* (Coloniae Agrippinae [Köln], 1550). However, for subsequent cartographers of the Holy Land, Ziegler's most important publication was his work *Quae intus continentur: Syria ad Ptolomaici operis rationem . . . Palestina . . . Arabia Petraea . . . Aegyptus . . . Schondia . . . Holmiae . . .; Regionum superiorum singulae tabulae geographicae* (Argentatori [Strassburg], 1532), republished in Strassburg in September 1536 under the title *Terrae Sanctae quam Palestinam nominant, Syriae, Arabiae, Aegypti & Schondiae doctissima descriptio, una cum singulis tabulis earundem regionum topographicis, authore Iacobo Zieglero Landauo Bavaro* , together with *Terrae Sanctae altera descriptio, authore Vuolffgango Vueissenburgio pridem Academiae Basiliensis Mathematico*. This 1536 volume also contains an *Index, totius operis locupletissimus, qui in priori editione desiderabatur*, and an *Elenchus, quo libro et capita Bibliorum et quoties singuli Palestinae loci continentur*, which gives biblical references to Ziegler's place-names. These two books (1532 and 1536 editions) are important for the history of the early mapping of Scandinavia (Schondia), but they are also extremely important for the development of Holy Land cartography.

Ziegler knew the work of Ptolemy, and his maps took their form and orientation from Ptolemy's *Quarta Asiae Tabula* (and not from the *Tabulae modernae*). The maps were oriented with north at the top. Degrees of latitude were printed on the left and right margins of the maps, and degrees of longitude along the top and the bottom. All sites for which Ziegler has given coordinates in his text can therefore be located, though not all of these appear on his maps. On his *Quinta tabula*, showing the whole of the Holy Land from Sidon in the north to Rhinocorura in the south, and from the coast on the west to Damascus in the east, a compass is shown in the top left-hand corner indicating the magnetic declination from true north. Ziegler is clearly anxious to present a work that meets the scholarly requirements of his time.

Ziegler's picture of Transjordan can be put together from several maps — the *Secunda tabula*, *Tercia tabula* (Fig. 11), *Quarta tabula*, *Quinta tabula* (Fig. 12), and *Septima tabula*; there is no single map covering the whole region. The *Quinta tabula* (the most commonly illustrated of the set) covers the land both sides of the Jordan from Syria and Mount Libanus in the north to the southern end of the Dead Sea, and the *Septima tabula* covers Egypt, Arabia Petraea, and the regions of the wilderness journey of the Israelites in the south. A fairly full picture of the Transjordanian regions can be obtained from these two maps. More detail of central and northern Transjordan can be obtained

FIGURE 11. Jacob Ziegler, Tabula Tercia (*Terrae Sanctae . . . descriptio*, 1536). (*The Board of Trinity College, Dublin*)

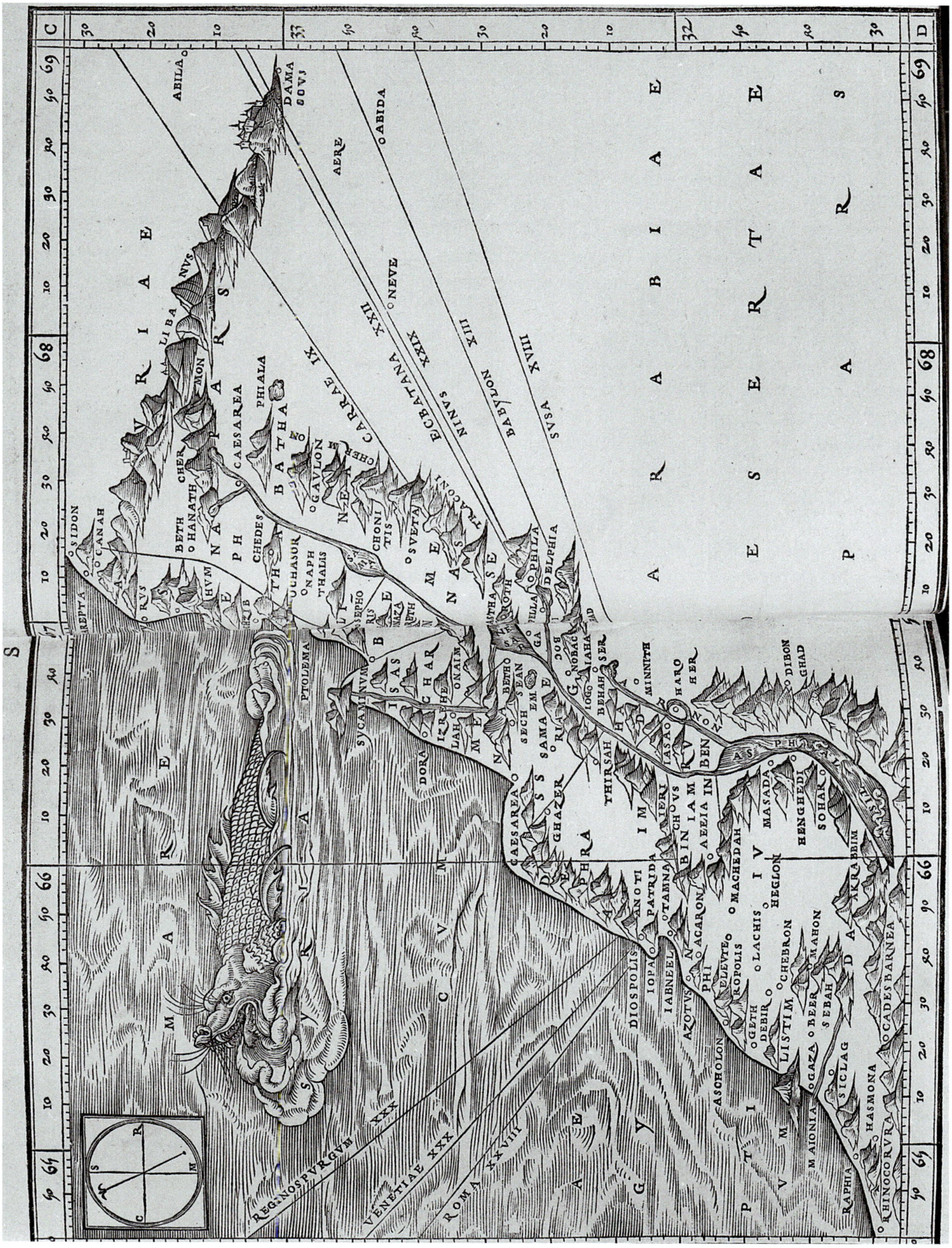

FIGURE 12. Jacob Ziegler, Tabula Quinta (*Terrae Sanctae . . . descriptio*, 1536). (*The Board of Trinity College, Dublin*)

from the other three maps. *Quarta tabula* shows the Dead Sea, with the Zered entering from the south-east and the Arnon and Jordan, enclosing part of Ruben's tribal land, from the north-eastern and north-western corners respectively; *Tercia tabula* shows the next section north, with the Arnon and Jordan running in parallel in a north-easterly direction, and the Jabboc entering the Jordan from the east shortly after the Jordan itself leaves the Lacus Genesar. It shows the land of the tribe of Gad lying mostly between the Jordan and the Arnon, and the land of the (half) tribe of Manasseh squashed in an acute angle between the Lacus Genesar and the Jabboc. The southern part of *Secunda tabula* overlaps with the northern part of *Tercia tabula*, but presents clearly the region east of the Jordan's upper reaches, as far as to its sources in Dan and Iar, and the Libanus mountain range, with Damascus away to the east. On these maps, physical features — seas, mountains, rivers — are drawn boldly, and names are printed in upper-case letters with a small ring indicating the position of the place named.

Ziegler tells us (1532 and 1536, f. xv^{v}) that he drew on the sacred history from Moses to the Maccabees, from Divus Hieronymus, Strabo, Josephus, Pliny, Ptolemaeus and Antoninus. He also used the accounts of Burchard of Mt Sion, and Bernard von Breitenbach. From Strabo Ziegler could have drawn little for the Transjordan region except references to Philadelphia, Machaerus and Aelana at the head of one of the recesses of the Arabian Gulf. Pliny similarly describes the Aelanitic Gulf, and mentions Machaerus and Callirhoe, the Lacus Asphaltites, the region of Peraea, and the Lacus Genesar surrounded by beautiful cities including on the east Julias (= Bethsaida) and Hippus (Ziegler [tercia tabula] shows Bethsaida west of the Jordan's entry to Lake Genesar; he does not show Hippus, locating it in his text [p. 49] 'super Iabboc fluvio'). Pliny usefully lists the towns of the Decapolis, including Damascus, Philadelphia, Gadara, Pella, Canatha, Panias/Caesarea, and Abila, all of which appear on Ziegler's maps. Pliny was clearly a useful source. Josephus would have given Ziegler information on such places as Coele-Syria (*Ant.* 1.145), the fountain of Phiala (*BJ* 3.509, 511, 513), Gaulon (*BJ* 4.2), Seleucia (*BJ* 4.2; *Vita* 187), Gamala (*BJ* 4.4–8), Ammanitis (*Ant.* 4.98; 5.262–63), Machaerus (*BJ* 7.171–89), Callirhoe (*BJ* 1.657), and Madian (*Ant.* 2.257). From the Antonine itinerary Ziegler could have drawn Abila, Damascus, Aere, Neve, Capitolias, and Gadara. With Ptolemaeus' map and approximately 60 names for the Transjordanian region, Ziegler's map shares Damascus, Samouli, Sueta, Iordanis fl., Lacus Genesar, Aspaltidis lacus, Capitolias, Gadara, Pella, Philadelphia, Livias, Callirhoe, Arabia Petraea, Petra, Elana, but ignores many others, particularly from the Arabian region, presumably because they were not immediately relevant to Ziegler's interest in the Holy Land.

Most of the names on these maps, however, are from the Old Testament. In his text, Ziegler takes each Israelite tribe in order — in Cisjordan, Asher, Naphthali, Zebulun, Isaschar [*sic*], Joseph (subdivided into Menasseh and Ephraim), Biniamin, Dan, Jehudah, Simon, and then for Transjordan, Bathanea et Thraconitis, Menasseh, Ghad, and Reuben — and names cities and other places belonging to each, using the tribal lists in Joshua as his starting point, and listing other names under the heading *extra autem Iehosuah*. Further sections are devoted to the wilderness journey from Egypt to Palestine, and to Arabia Petraea (pp. lxxvi–lxxxiiii). For the cities of the Transjordanian tribes Ziegler gives 59 sets of coordinates, for the Israelite wanderings in the wilderness 28, and for Arabia Petraea 48, but in each case he names a number of sites — about 35 for the Transjordanian tribes, 12 for the wanderings, and about 25 for Arabia Petraea — for which he gives no coordinates. If we list the Transjordanian city names (ignoring names of tribes, regions, natural features, and repetition from one *tabula* to another) which appear on Tabulae 2, 3, 4, and 5, we find the following:

Salcha	*Caesarea*	Magedon	Seleucia	Neve	Aere
Capitoliada	Samuli	Sueta	*Argob*	*Chorazin*	*Astaroth*
Gamala	Amathus	Gadara	Abila	*Astaroth*	Pella
Iabis	*Philadelphia*	*Penuel*	*Nobach*	*Saphon*	*Iogbehah*

Canatha	*Medehar*	*Iahazer*	*Bethnimrah*	*Machane*	*Betharan*
Debir	*Chesbon*	*Haroher**	Beasthra	*Minnith*	*Medebah*
Iahasah	Libyas	*Abilah***	Machrus	Callirhoe	*Bethhaiesmoth*
Bosor	*Dibon Ghad*				

Biblical names italicised
* Haroher in the middle of Arnon
** Abel haShittim

To these we may add a number of Transjordanian names from *Septima tabula*, which covers Egypt, Sinai and Arabia Petraea. Most of the names on this map come directly from the list of halting-places (*mansiones*) on the Israelite march from Egypt to the Promised Land in Num. 33 (his predecessor Cranach, however, used the story-line of Exodus-Numbers as his chief source). Of the 40 halting-places from Rameses to Almon Diblathaim, Ziegler locates 28, giving no coordinates or place on his map for Sina, Chibroth Hattahavah, Rithma, Rimmon peres, Libna, Rissah, Cheleloth, Cheredah, Macheloth, Thahath, Tarach, Mithchah, Casmonah and Ben Iahachan, though mentioning them in his text. Apart from these *mansiones*, we find on this map

Albus vicus	*Bamoth vallis*	*Beeroth*	Colchiel	*Dedan*	*Disahab*
Elana	*Haroher*	*Cademoth*	*Luchit*	*Mathanah solitudo*	*Madian*
Nachaliel		Petra regia	*Pharan*	*Puteus* [= Beer, Num. 11.16]	*Theman*

Ziegler has combined the classical and biblical sources, but has made fuller use of the biblical sources than his predecessors. He has located on his maps only sites for which he can give coordinates (and not all of those), and it is particular noticeable that he has not attempted to identify places by the illustrations of the biblical stories attached to them, as in earlier maps. The *sepulchrum Iob*, the *terra Hus*, the various descriptions of the Dead Sea and its drowned cities, the story of Balaam, and the site of John's baptismal activity have all disappeared, as have also the reference to the treasure-burying activities of the sultan (*soldanus*), and the Crusaders' castles at Petra deserti and Mons regalis. The only reference on these maps to a biblical story is the *igniti serpentes* located in 'Petraea Arabiae desertae pars', complete with the picture of a serpent mounted on a pole, on *Septima tabula*, whose title refers to *itinera Hebraeorum per desertum*. Ziegler is concerned with the biblical world, but his detached and academic presentation of the subject follows the style of the Ptolemaic *Quarta Asiae tabula*, not the approach of the *Tabula moderna terrae sanctae*. Ziegler does not follow the contents of the *Tabulae modernae*; he goes firmly to the biblical sources for his place names. (Colchiel is a puzzle: it is not a biblical or classical place-name. Ziegler's text associates it with the *metalla*, 'mines', locating it 11 miles from Katachrysea or Disahab [Dt. 1.1]; his map locates it between Disahab and Punon, both associated with mines [Euseb. *Onom.* 114]. The name perhaps derives from the *chalkou metalla*, 'copper mines', of Euseb., *HE* VIII.13.5; cf. Migne, *PG* XXV (1857), 766.) One notes also that Ziegler has shaken free of the confusion that located Idumaea next to Syria and Damascus. Chorozain, however, remains stubbornly east of the Sea of Galilee, though Engedi reappears correctly west of the Dead Sea, even if wrongly located south of Masada. The Arnon river enters the Dead Sea too far north and from too north-easterly an angle, paralleling the Jordan. The Zered appears more accurately placed. The Dead Sea is banana-shaped, concave to the west, setting a fashion for years to come. The Elanitic Gulf is correctly separated from the western arm of the Red Sea.

There can be little doubt that true credit for the scholarship behind the well-known sixteenth-century maps of the Holy Land and Transjordan — those of Mercator, Tilemann Stella and others — belongs to Jacob Ziegler. Ziegler combined classical and biblical scholarship of a high order — he clearly had first-hand knowledge of the ancient sources — with a strong grasp of the geographical science of his day. He published a learned, detailed volume (1532, 1536), whose value was instantly recognised and proclaimed by the young Mercator in his map of 1537. Ziegler's work, in fact, is the real basis for all maps of the Holy Land and Transjordan until the beginning of serious topographical research on the ground in the eighteenth and nineteenth centuries. He deserves more scholarly recognition than he has hitherto received.

CHAPTER FIVE

THE SIXTEENTH-CENTURY CARTOGRAPHERS

MERCATOR

THE SECOND EDITION of Ziegler's work was published at Strassburg in 1536. For all the crudity of the woodcuts, Ziegler's maps were based on sound scholarship, and it is no surprise that his work was immediately used by an important successor. In 1537, a young man named Gerardus Mercator (1512–94) presented his map of the Holy Land under the title *Amplissima Terrae Sanctae Descriptio ad Utriusque Testamenti Intelligentiam.* This map (Fig. 13) was published at Louvain by copperplate engraving in six sheets, designed to be pasted together in order to form a large wall-chart 17 inches high by 38¼ inches long (43.4 × 98.4 cm). The only known surviving copy of the original publication was found in the public library of Perugia [I.C.94 (3)] and published by R. Almagià (1927); in the early seventeenth century a new impression was made from the original plates by C. J. Visscher in Amsterdam (see Nebenzahl 1986, 72–73; for a full description, see Delano-Smith and Ingram 1994, 268–83). Mercator offered a new view of the Holy Land, drawn as if by an observer looking down from above the Arabian desert in the south-east towards the Mediterranean Sea; the Egyptian-Palestinian coastline is shown running in a convex curve across the top of the map, with the Nile Delta to the left and Phoenicia on the top right. The Red Sea (Pars Sinus Arabici) appears bottom left. In conception, however, the map is not unlike the map of the promised land that appears (inverted) in the Coverdale Bible of 1535, which also draws on Ziegler's work (see the facsimile of *The Coverdale Bible 1535* [1975], p. 48; Delano-Smith and Ingram 1991, Fig. 18, pp. 40–41). The centre is given over entirely to the Israelites' march through the wilderness from Egypt to Canaan, as in the maps of Cranach and Coverdale. While Cranach used the biblical Exodus and Numbers narratives, Coverdale used the Num. 33 list, and Mercator used Ziegler's coordinates of the 42 *mansiones* of Num. 33 to identify and mark the route (carefully listing without locating those *mansiones* for which Ziegler supplied no coordinates). Ziegler's location of halting-places nos 27–30 to the west caused Mercator to turn the Israelites' route in a westward loop, circling back to cross the former line of march and proceed eastwards round the southern end of the Dead Sea to Moab. This route located Hesion-chaber (Ezion-geber) (no. 32) between Hebrona (no. 31) and the desert of Sin and Cades (no. 33), in the middle of the wilderness. This had the advantage that the route did not have to double back (as in most later maps) to Ezion-geber at its biblical (1 Kings 9.26) location on the Gulf of 'Aqaba. Ziegler presents us with two Ezion-gebers, one in the wilderness and the other, the port, on the Gulf of 'Aqaba. Mercator dealt with this problem by ignoring the Gulf of 'Aqaba, Elath, and the nearby Ezion-geber, on his map. In a map of Europe published in 1554 (revised 1572 [Fig. 14]; on this map, see Dürst in Watelet 1998, 31–41), however, Mercator revises the route through the wilderness, and instead of looping the Israelites to the west he makes them first travel east-north-east to Sinai, then north to Asemona, then sharply south past Beeroth to Ietabatha, where they turn north again past Ebrona, Asiongaber, Sin Desertum and past Phunon to Dibon Gad and finally west to Jericho. Interestingly, on this map Mercator recognises the Elanitic Gulf (cf. also on his *Map of the World*, 1669 [van 't Hoff 1961, sheets 10 and 11]) and marks Elana at its head, with Hesiongaber further south on the west coast of

FIGURE 13. G. Mercator, map of Palestine, 1537 (Perugia, Biblioteca Communale Augusta, I.C.94 [3]). (*Biblioteca Communale Augusta, Perugia*)

FIGURE 14. G. Mercator, *Europae tabula*, 1572 (revised from *Europae descriptio* of 1554) (Perugia, Biblioteca Communale Augusta, I.C.94 [2]). (*Biblioteca Communale Augusta, Perugia*)

the Gulf, but he does not show the Israelites visiting it (cf. Delano-Smith & Ingram 1994, 279–80). In the large cartouche on the lower right-hand corner of his 1537 map, Mercator addresses the reader: 'Candido lectori s[alus]. Palestinam hanc, & in eam per Arabicas petras ex Aegypto Hebreorum iter ex Zieglero fidissimo horum chorographo deprompsimus, nominibus illius ad veterem Bibliorum translationem, quantum potuim, variatis, quando huic ut usitatissime potius inserviendum videretur.' ('Fair reader, greeting! We have drawn this map of Palestine, and the Hebrews' route into it from Egypt through the stony regions of Arabia, from Ziegler, the most faithful cartographer of these things, his place-names being adjusted, as far as possible, to the old translation of the Bible, since this seemed most serviceable for our purpose.') Mercator states expressly that he has drawn his map of Palestine and of the Israelites' route from Egypt from Ziegler, 'the most faithful cartographer of these things'. (He notes also that for his biblical text he drew on the work of Santes Pagninus, the Dominican biblical scholar (1470–1536), whose translation of the Bible was published at Lyons in 1528.) Clearly, Mercator was chiefly concerned to illustrate the route of the Israelites from Egypt (cf. Cranach; for this theme in maps in sixteenth-century Bibles, see Delano-Smith and Ingram 1991). This theme is emphasised by the vignettes of the Israelite camp round the Tabernacle in the desert (cf. Num. 2), the battle of the Amalekites and Israel at Rephidim (Exod. 17), and the serpent in the wilderness (cf. Num. 21.8) near Salmone, and in particular by the quotations from Micah 6.3-4 ('I brought you up from the land of Egypt ...') and Deut. 8.7-10 ('the Lord your God is bringing you into a good land ...') set below the scene of Christ on the cross at the top centre of the map.

Given that the whole of both Cisjordan and Transjordan are presented together on one of the six sheets, the upper right-hand sheet of the set, it is surprising that Mercator has managed to include over eighty Transjordanian names. A number of these are familiar from Ptolemy: for example, Lacus Tiberiadis, Sueta, Pella, Capitoliada, Samouli, Canatha, Callirhoe, Arabia Petraea, Petra, Sinus Arabici pars. Almost all the rest appear in Ziegler's text or on Ziegler's maps (though the spellings of the place-names conform to Pagninus, not Ziegler). Clear indications of direct borrowing are the references to Colchiel (Ziegler 1532, 83) near Petra deserti and Theman, to Telemanum (Ziegler 1532, 52), and to Medebar and Medaba in the region of Gad. (Telemanum is another curiosity, neither biblical nor classical; Ziegler found the name in Pliny, and located the place 'media distantia inter Damascum et Petram'; cf. Pliny's [*NH* v.89] 'Elatium, dimidio propior Petrae quam Damascus'. Is Telemanum, similarly located by reference to Damascus and Petra, a confused or half-remembered Elatium? After Mercator and Tilemann cartographers omitted it.) Mercator, however, has selected from places for which Ziegler has given coordinates, and he derives his material from well-known passages from both the Old Testament (including the books of Maccabees) and the New Testament (as the title of his map indicates). From the New Testament he draws on the story of the Gerasene/Gadarene swine (Mk. 5), Jesus' utterance against Chorazin and Bethsaida (Mt. 1.21), both taken as located east of the Sea of Galilee, and Luke's references to Iturea and Traconitis (Lk. 3.1). One vignette illustrates the story of Jesus' Transfiguration on Mt Tabor. From Judas Maccabaeus' wars in Transjordan Mercator locates Bosor (1 Macc. 5.26) and Ephron (1 Macc. 5.46, 2 Macc. 12.27). From the Old Testament, Mercator draws particularly on the stories of Jacob in Transjordan (Gen. 32), of the Israelites in Transjordan (Num. 21, 32–33), of the tribal allocations east of the Jordan (Josh. 13), and of the campaigns of Gideon and Jephthah east of the Jordan (Jud. 8, 11). That is, Mercator is in fact illustrating some well-known stories of the settlement of Israel, which underlines the point that his main focus is the wilderness route from Egypt to Canaan.

It is hard to know Mercator's precise intentions. His title suggests that he is concerned with a better understanding of both Testaments, New and Old. Nicholas Crane (Crane 2002, 91) has suggested that Mercator's composition, with its crucified Christ at top-centre, may have been modelled on a recent woodcut by Cranach depicting the Lutheran opposition of Law and Gospel. If so, Mercator is strongly emphasising the gospel at the expense of the Law, for there is no vignette illustrating Moses' reception of the Law on Mt Sinai, but there is a small vignette of the transfigured Christ (at his transfiguration, Christ appears centrally with Moses and Elijah on either side; cf. Mk. 9.4). The approving reference to Ziegler is revealing, for Ziegler had important connections with men like Erasmus and Luther, and had taken the Protestant side in the contemporary struggles (later, however, reverting to his native Catholicism).

WOLFGANG WISSENBURG

Wolfgang Wissenburg (1496–1575) was born about 25 years after Ziegler and about 16 years before Mercator. He was a biblical scholar from Basel, and he contributed a biblical gazetteer entitled *Terrae sanctae Descriptio* to Ziegler's 1536 volume. In 1538 he published at Basel a wall-map, *Descriptio Palestinae Nova* (Nebenzahl 1986, 74–75), of 8 sheets (the whole being 41 × 29 in., 105 × 74.5 cm), which he dedicated to the Archbishop of Canterbury, Thomas Cranmer, presumably in approval of Cranmer's reformationist views. Cranmer had just supported the publication of Matthew's Bible (1537), an English translation largely based on the work of Tyndale (Daniell 2003, 194). Wissenburg's 1538 map, *Descriptio Palaestinae Nova*, of which one copy survives in the Bibliothèque Nationale in Paris, is largely dependent on Ziegler, presenting a similar coastline, and incorporating a compass rose and straight lines indicating distances from Rome, Corinth, Thessalonica, Ephesus, Cyprus, Charran, Ninive, Babylon, Ecbatana, Susan, and Persia. Like Mercator, Wissenburg follows Num. 33 for the route of the Israelites' wilderness march, but lists and numbers all the *mansiones* in sequence on a clearly defined path (like Cranach, Wissenburg seems to focus on the Israelites' wilderness march, which takes up the lower centre of the map, with Sinai prominent in the middle). He locates Aezion Gaber at the head of Elaniticus Sinus, clearly distinguishing this from the Heroopolites Sinus (Gulf of Suez) at the northern end of the Red Sea. The Dead Sea is straighter, and less banana-shaped than Ziegler's. In the Transjordan region, the river Zered is not shown; the Arnon enters the Dead Sea at its northern end from the south-east (and not in parallel with the Jordan as on Ziegler's map), and the Iaboc enters the Jordan near Bethshan, just before the Jordan reaches the Lacus Genesar (Sea of Galilee) (as indeed on Ziegler's map). Apart from three mountains (Libanus mons, Seir mons, Montana Felicis Arabiae), two rivers (*torrens*) (Iaboc, Arnon), a lake (Genesar), a sea (Mare Mortuum) and a gulf (Elaniticus sinus), and the regions of Traconitis, Iturea, Perea, Idumea, the tribes of Ruben and Gad, and the campestria Moab, Wissenburg's map locates a mere handful of cities (Damascus, Gerasa, Debir, Mahanaim, Rabbath Ammon, Bethabara, Medba, Esbon, Aroer, Iahatz, Sodom, Gomorra, Petra regia, and Elath, with Rhabmathom and Mesada in the desert to the east). Wissenburg has selected his place-names from Ptolemy and Ziegler and perhaps from Mercator's 1537 map. He seems to have been fairly uninterested in the Transjordanian region; apart from events on the wilderness march, he shows no vignettes of any biblical events taking place in the Transjordan, and though he had plenty of space available on his map he omits many Transjordanian place-names which appear in Ziegler's work. That he knew of more Transjordanian places than he mapped is clear from his *Descriptio* of 1536, which includes the biblical place-names Ar, Phiala, Hippos mons, Hutz terra, Gadera, Gilead, Havoth Iar, Matanah, Nahaliel, Hor, Pharan, and Zered.

SEBASTIAN MÜNSTER

An unusual map of the Holy Land (*Terra Sancta. XIII. Nova tabula*) was published by Sebastian Münster (1489–1552) in his 1540 edition of Ptolemy's *Geographia*. This map ([1552 ed.] Fig. 15) presents the Holy Land from the east, or, more precisely, the north-east, with south-west at the top of the map and north-east at the bottom. The Holy Land is viewed from the top of Mons Hermon, with Phoenicia away to the right, Arabia Petraea to the left, and with the kingdom of Basan, Traconitis, and the Gessurites and Ammonites at one's feet, immediately to the south. Caesarea Philippi and the sources of the Jordan appear on the lower right of the map; the Jordan passes through Lacus Samachonitis (unnamed), and diagonally

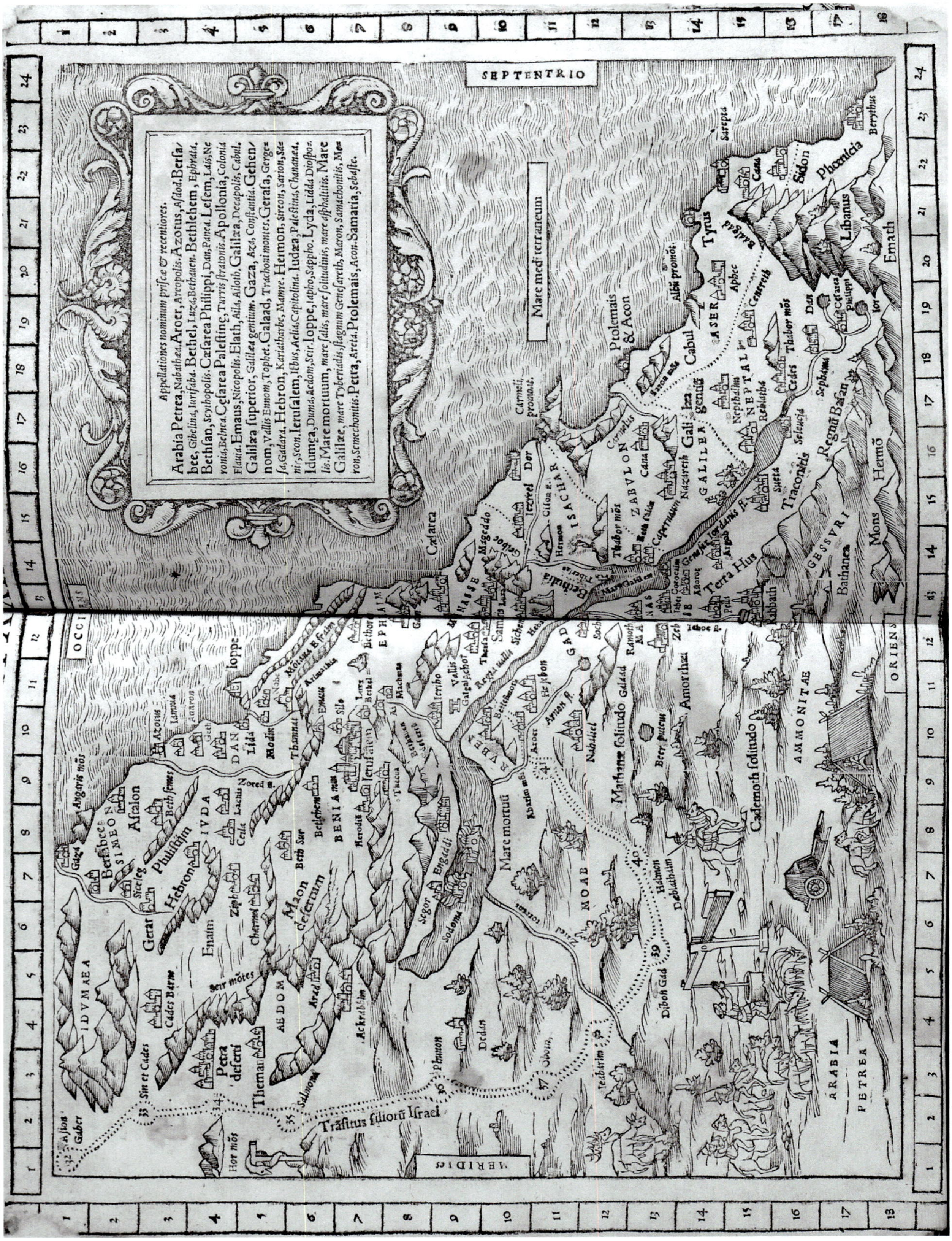

FIGURE 15. Sebastian Münster, Terra Sancta. XIII. Nova Tabula (*Geographia* 1552 [1540]). (*The Board of Trinity College, Dublin*)

across the map towards the Sea of Galilee and ultimately to the Dead Sea in the middle distance. Parallel with the Jordan the Arnon flows into the north-east corner of the Dead Sea; the Zered flows into the middle of the Dead Sea's east shore. Asion Gaber appears in the very top left-hand corner of the map; from there the Israelites' track through the wilderness and Moab is shown, via Sin et Cades, Hor mons, Salmona, Phunon, Oboth, Ieabarim, Dibon Gad, Halmon Daeblathaim, and Abarim montes. The tribal possessions of Ruben, Gad and (half-)Manasse are outlined.

A comparison of Münster's map with those of Wissenburg (1538), Mercator (1537), and Ziegler (1536) shows that Münster was not using Wissenburg's map, for his list of place-names has relatively few in common with Wissenburg and relates much more closely to the list of those used by Mercator (1537) (note especially Zeb, which Mercator had from Ziegler). Münster was probably using Ziegler selectively (note the inclusion of Sephama, Seleucia, Bathanaea, Regnum Basan, Terra Hus and Gessuri, which do not appear on Mercator's map).

TILEMANN STOLZ (STELLA) AND ORTELIUS

In 1552 Tilemann Stolz (Latinised to Stella) published at Wittenberg his first map of Palestine, entitled 'Chorographic map of the celebrated places in the Kingdom of Judea and the Kingdom of Israel, for the understanding of the sacred scriptures, etc.' (Karrow 1993, 500). Though 600 copies were printed, no known survivor exists. However, this map was used by de Jode in his *Speculum orbis terrarum* (1578) and by Ortelius in his *Parergon* (1586–1624) map 3, dated 1586. Ortelius entitles the map *Typus chorographicus, celebrium locorum in regno Iudae et Israhel, arte factus a Tilemanno Stella Sigensi*, noting in an oval cartouche that 'Abrahamus Ortelius in hanc formam minorem redigebat. Anno MDLXXXVI' ('Abraham Ortelius reduced it to this smaller format. Year 1586'). The reference to 'this smaller format' suggests that the original was larger. Tilemann published a second map at Wittenberg in 1557 focusing on the Exodus and entitled *Itinera Israelitarum ex Aegypto loca et insignia miracula diversorum locorum et patefactionum divinarum descripta a Tilemanno Stella Sigenensi ut lectio librorum propheticorum sit illustrior* ('The journeys of the Israelites from Egypt, the places and famous miracles of different places and divine revelations described by Tilemann Stella of Siegen that the reading of the prophetic books might be made clearer'). This was a woodcut map, on nine sheets, which perhaps suggests that the earlier map was also printed on several sheets (hence Ortelius' 'forma minor'). Only one copy is known to exist (Karrow 1993, 502). Two years later, in 1559, Bernard Puteanus in Antwerp published the same map, on nine sheets; this has been reproduced in smaller size in Nebenzahl (1986, 76–77). This map was used by Ortelius in his *Theatrum orbis terrarum* (1570) under the title *Palestinae sive totius Terrae Promissionis nova descriptio auctore Tilemanno Stella Sigenens*. It reappeared, printed from a recut plate, in Ortelius' *Theatrum* of 1579 (the version in the Library of Trinity College, Dublin [PP.a.34], bears a date, 1579, in a small cartouche above the main cartouche), and yet again from a newly engraved plate in Ortelius' *Parergon* (1595–1624) (this version has the biblical reference 'Deut. 8' in the upper small cartouche, and the Latin text of Deut. 8.7-9 in the larger cartouche below [see Fig. 18]). Ortelius' 1570 version was the immediate source for the English map of Canaan/Palestine engraved by 'Humfray Cole Goldsmith a English man born in ye north and pertaining to ye Mint in the Tower 1572' (see Barber in Ackermann 1998, 97–100). We have therefore plenty of evidence for the contents of Tilemann Stella's two maps of Palestine, the first illustrating the kingdoms of Judah and Israel, and the second the Israelites' journey through the wilderness to the promised land.

The orientation of Stella's first map of Palestine is not entirely certain, for de Iode's 1578 version (Fig. 16) presents the Palestinian coastline running along the top of the map, with north on the right and south on the left (similar to that used by Mercator's 1537 map), while Ortelius' 1586 version (Fig. 17) shows the coast running diagonally across the map, with 'Phoeniciae pars' in the top right hand corner and 'Aegypti pars' in the bottom left. This is the orientation used by Stella in his second map of 1557/1559.

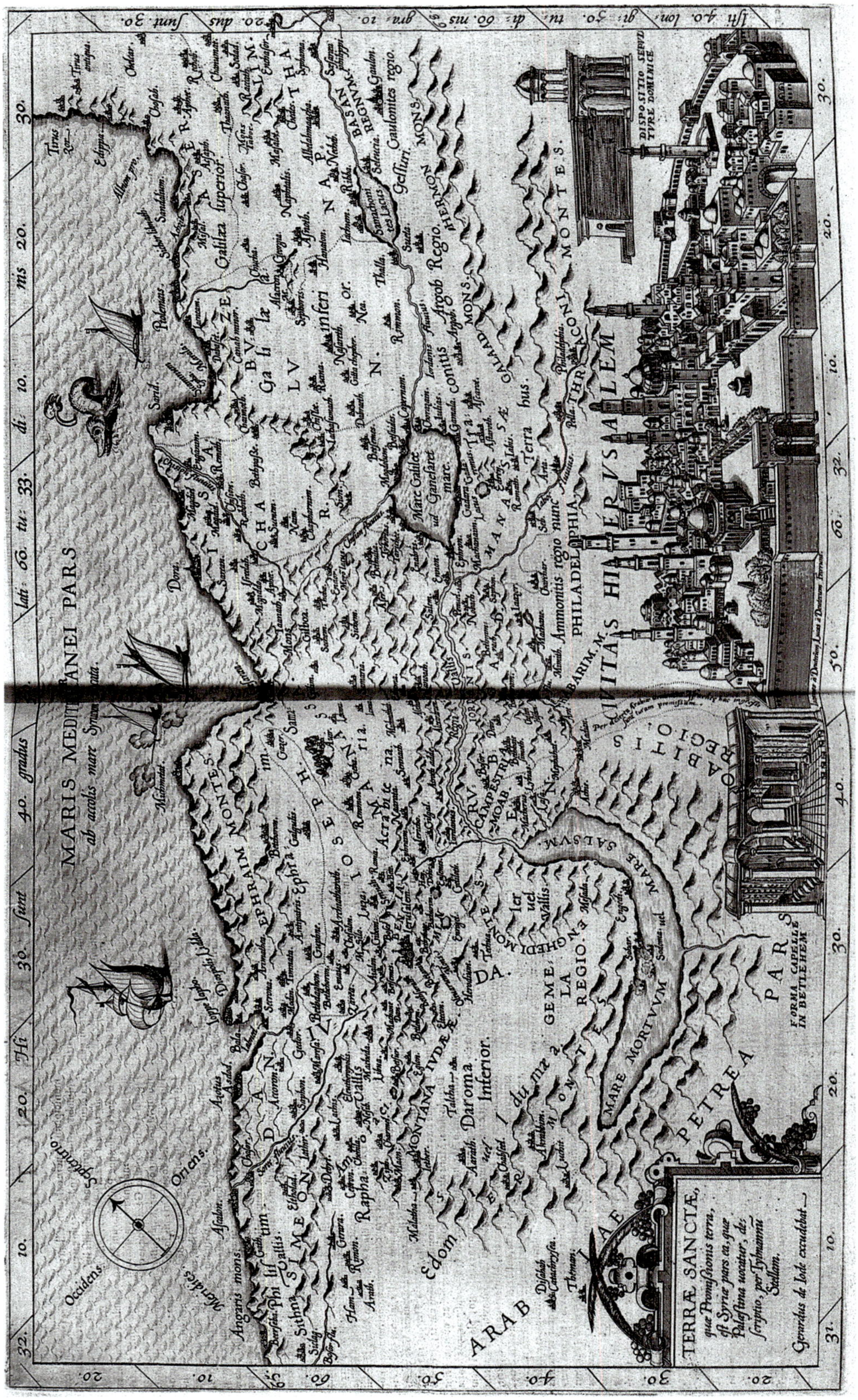

FIGURE 16. Gerardus de Jode, 'Terrae Sanctae, quae Promissionis terra, . . . descriptio' (*Speculum orbis terrarum*, 1578). This is de Jode's version of Stella's earlier (1552) map. (*The Board of Trinity College, Dublin*)

FIGURE 17. Tilemannus Stella, 'Typus chorographicus, celebrium locorum in regno Iudae et Israhel, arte factus a Tilemanno Stella Sigenensi', from A. Ortelius, *Theatrum Orbis Terrarum parergon* [1586–1624; ed. 1624], map iii. This map is Ortelius' version of Stella's earlier (1552) map. (*The Board of Trinity College, Dublin*)

Whether Stella followed Mercator's 1537 orientation or not for his first map, it is clear that he used Mercator for his general outlines — his Dead Sea is banana-shaped, with the Zered running into the south-east shore, the Jordan and the Arnon running parallel into the northern end of the Dead Sea, and the Jaboc running into the Jordan roughly along the course of the present Yarmuk — and for the place-names. His list of Transjordanian place-names, to judge from Ortelius' 1586 version, is almost exactly that of Mercator's 1537 map. Ortelius' 1586 version, however, gives more Transjordanian names than de Iode's 1578 version; de Iode adds Terra Hus and Seleucia, but omits over 20 names (Gileaditis, Abila Vinearum, Amathus, Betonim, Mispeh, Capitolias, Samuli, Medebar, Medba, Jogbehah, Canatha, Beasthra, Telemannum, Damnaba, Nebo mons, Asdothpisgah, Bethbaalmeon, Nachaliel, Beer Puteus, Mathanah solitudo, Cademoth solitudo, and Colchiel). Again, it is not entirely clear whether it was Stella who omitted these names in drawing on Mercator's 1537 map, or de Iode who omitted them in drawing on Stella's first map. As some of these names (Capitolias, Samuli, Asdoth [Pisgah], and Colchiel) appear on Stella's second map, Stella clearly knew them, and it seems likely that Stella's first map did indeed follow Mercator closely, and that it was de Iode who abbreviated the list of Transjordanian names from Stella's first map.

Stella's second map (1557/59) (Fig. 18) was designed to focus on the Exodus and the Israelites' wilderness journey. It therefore includes the southern, desert region of Palestine towards the Red Sea and Egypt, and, like Mercator's 1537 map, illustrates the route of the Israelites from Egypt to the Promised Land as given by Num. 33. It shows only one prong of the Red Sea, that leading towards modern Suez, and does not indicate the Gulf of 'Aqaba, though placing Asion Gaber at the southernmost point of the Israelites' route, near the shore of the Gulf of 'Aqaba. This map, however, offers only a selection of the Transjordanian place-names given on the 1552 map and its source, Mercator's 1537 map. Among those omitted is Telemannum in the Syrian desert, though Colchiel south of the Dead Sea remains. However, it adds Pella in northern Transjordan, and includes with Petra a list of places identified with it (Sela, Arachen, Recem, Iathkeel, Nabath).

PETRUS LAICKSTEEN AND CHRISTIANUS SGROOTEN

While Stella was working on his maps, a cartographer named Petrus Laicksteen was travelling round the Holy Land (1556), acquiring personal knowledge of the geography and topography, and making notes. He had already drawn a plan of Jerusalem (1544); on his return from Palestine he seems to have passed on his notes to a professional cartographer in the employ of Philip II of Spain named Christianus Sgrooten (*c.* 1532–1608; originally from Sonsbeck in what is now Germany). The Laicksteen-Sgrooten map was published by Hieronymus Cocus (Cocq) at Antwerp in 1570. The map was printed from copper plates in nine sheets, as Stella's *Itinera Israelitorum ex Aegypto* map had been in 1559. A text at the top of the map gives its origins: 'Nova descriptio amplissima Sanctae Terrae quam M[agiste]r Petrus Laicksteen Astronomus perambulavit ac visitavit An[n]o rede[mp]tionis 1556 per Christianum Sgroothenum Reg[is] Ma[iesta]tis Hispan[iae] etc. Geographum collecta 1570' ('A new, most splendid description of the Holy Land which Master Petrus Laicksteen, Astronomer, toured and visited in the Year of Redemption 1556, put together by Christian Sgroothen, Geographer of his Majesty the King of Spain, 1570') (Nebenzahl 1986, 82–83, pl. 29). In 1584 Ortelius printed (*Addimentum III, Theatrum Orbis Terrarum*, fol. 12) a map 'Terra Sancta a Petro Laicstein perlustrata et ab eius ore et schedia . . . redacta'.

The 9-sheet spread contains two maps, both oriented with north at the top: the larger map extends from Mount Hermon in the north to the southern end of Dead Sea and shows the territories of the tribes of Israel, while the smaller map, set into the top left of the whole spread, focuses on the Exodus and shows the land both sides of the Jordan and Dead Sea from Egypt and the Red Sea in the south to Mons Libanus in the north. This double focus, on the Exodus-Wilderness story and on the tribal territories or the kingdoms of Judah and Israel, was becoming well

FIGURE 18. Tilemannus Stella, 'Palestinae sive totius terrae promissionis nova descriptio auctore Tilemanno Stella Sigenensi'. This is Ortelius' version of Stella's later (1557) map. Earlier versions of this map appear in the *Theatrum Orbis Terrarum* (1570, 1589). This version, from the *Parergon* (1595–1624), gives the text of Deut. 8.7–9 in the cartouche, describing the Promised Land.
(*The Board of Trinity College, Dublin*)

established among cartographers of the Holy Land. Cranach, Mercator and Stella had already presented the wilderness journeying of the Israelites, and Laicksteen was following a well worn trail. The portrayal of the Dead Sea as turning east at the southern end may possibly suggest that Laicksteen had observed — from the west side of the Dead Sea — the southern shore of the Lisan peninsula, and the suggestive curve on the northern shore of the Red Sea may possibly indicate that Laicksteen knew of the Gulf of 'Aqaba; but if he did, he does not relate Asion Gaber to it, locating this well to the north in the sequence of the *mansions* of the wilderness march. Laicksteen's *Arnon torrens* enters the north-east corner of the Dead Sea after uniting with a *Dibon torrens*, so he was not familiar with the east side of the Dead Sea. However, he has the *Iaboc torrens* entering the Jordan mid-way between the Dead Sea and the Sea of Galilee, as it does in fact, whereas earlier maps showed it entering the Jordan nearer the Sea of Galilee.

Of particular interest is the list of Laicksteen's Transjordanian place-names. What is interesting is that Laicksteen does not appear to use or follow Stella's list. On the larger map, only about 20 out of about 130 names correspond with names on Stella's maps, and those that do correspond tend to differ in their spelling. Laicksteen seems to have drawn eclectically on the Bible, using Maccabees (cf. Torrens Datheman, which runs into the south-east coast of the Sea of Galilee, 1 Macc. 5.9) as well as the Old Testament and the Gospels. Ortelius clearly approved of this map, using it, with due acknowledgement, in his 1584 *Addimentum* III (fol. 12) of his *Theatrum Orbis Terrarum*. For this work, he took the larger of Laicksteen's two maps, and re-oriented it so that east lay at the top. He used Laicksteen place-names, copying them almost exactly. He followed Laicksteen's presentation of the coasts, inland seas and rivers, though as he was using the larger map he did not show the Dead Sea with its revised south-east corner as given on the Exodus-wilderness route map. This map also appeared in the 1606 edition of Mercator's (Amsterdam: I. Hondius) under the title *Terra Sancta quae in Sacris Terra Promissionis ol[im]. Palaestina* (Fig. 22).

ARIAS MONTANUS

At the end of the text accompanying Stella's second map, Ortelius notes 'De veteri Palaestina legendi sunt quoque D. Hieronymus, & ea quae scripsit nostro tempore B. Arias Montanus in suo Chaleb'. Arias (1527–98) was a Spanish linguist and theologian, best known for his work in supervising the publication of the Antwerp Polyglot Bible (8 vols, 1568–73). In the final volume (*Phaleg, sive, De Gentium Sedibus primis, orbisque terrae situ, Liber*. Antverpiae MDLXXII) Arias presented theological essays and included four maps and a plan: (1) *Orbis tabula*; (2) *Tabula terrae Canaan Abrahae tempore et ante adventum filior[um]. Israel cum vicinis et finitimis regionibus*; (3) *Terrae Israel omnis ante Canaan dictae in tribus undecim distributae accuratissime et ad sacras historias intelligendas oportuniss[ime]. cum vicinarum gentium adscriptione tabula et exactissimo mansionum XLIII situ*; (4) *Descriptio et situs terrae sanctae alto nomine Palestina multis praeclaris historiis et miraculis utriusque testamenti a Christo Salvatore*; (5) *Antiquae Ierusalem vera iconographia ad sacrae lectionis praecipue et aliarum de illa urbe historiar[um] explicationem*.

Arias offers three maps of Israel. The first (*Tabula terrae Canaan Abrahae tempore*) (Fig. 19) claims to show the land of Canaan in Abraham's time, before the arrival of the Israelites, and the second (*Terrae Israel omnis ante Canaan dictae . . . tabula*) (Fig. 20) the land of Israel distributed between the eleven landed tribes. The third (*Descriptio et situs terrae sanctae . . .*) locates various historical and miraculous events from the Old and New Testaments. The most obvious thing about these maps is that their outline goes back beyond Ortelius, Stella and Ziegler to the format of Sanutus and the late fifteenth-century 'modern' maps of Palestine, oriented with east at the top and the Palestinian coast running from left to right along the bottom of the map. On the left, i.e., to the north, a mountain chain running west-east divides Syria from the Holy Land, and another mountain chain running north-south along the top of the map closes off Arabia to the east. The Holy Land thus appears as a rectangle, with the Jordan valley running north-south across the middle, and the Red Sea (with no sign of the Gulf of Aqaba) at the southern end. The Jabbok (Iaboc) river enters the Jordan just south of the Sea of

FIGURE 19. Benedictus Arias Montanus, 'Tabula Terrae Canaan Abrahae tempore' (*Phaleg* 1572). (*The Board of Trinity College, Dublin*)

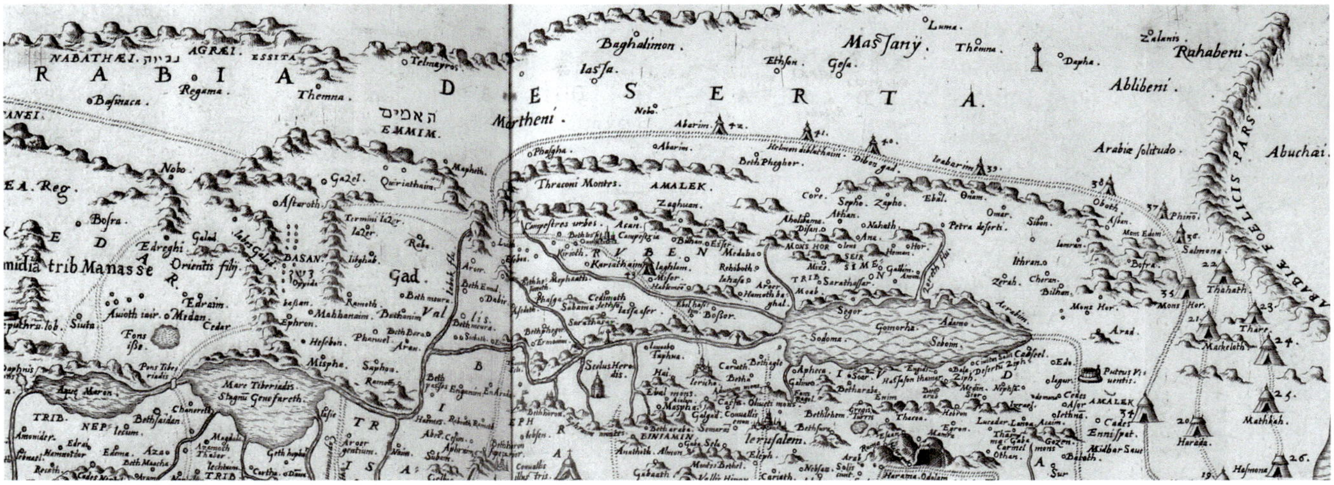

FIGURE 20. Benedictus Arias Montanus, 'Terrae Israel omnis ante Canaan dictae in tribus undecim distributae ... tabula' (*Phaleg* 1572). (*The Board of Trinity College, Dublin*)

Genesareth; the Jordan enters the north-west corner of the Dead Sea, with two other unnamed rivers entering the Dead Sea in the north-east corner. On the map of Canaan in Abraham's time, the river Arnon enters the Dead Sea from the east, two-thirds of the way down the coast (not entirely inaccurately); however, on the map showing Israel's tribes, this river has become the Zoreth flu., and the Arnon is perhaps the river entering the north-east corner of the Dead Sea, as in Ziegler's and Stella's maps. Arias' source is not clear. He claims to have 'obtained a map of Canaan in Trent from a learned Mantuan, very knowledgeable in Hebrew, who had commissioned it at great cost' (Shalev 2003, 63). Shalev believes that 'Montano's own map of Canaan is based on the Tridentine map, which he annotated and augmented with a descriptive text in order to facilitate the understanding of biblical toponymy' (ibid., 63). 'Montano may be referring to the anonymous Mantuan Hebrew map of *c.* 1560, which was rediscovered in 1991 in Zurich' (ibid. 77, n. 47), reproduced in Tishby 2001, 121 [not 127, as given by Shalev].

Each of these two maps (the 'Tabula terrae Canaan Abrahae tempore' and the 'Terrae Israel omnis ante Canaan dictae . . . tabula') give about 120 place-names for the Transjordanian region, of which about 65 are held in common. For the place-names on the map of Canaan in Abraham's time, Arias drew almost entirely on the Pentateuch, using especially the Table of Nations (Gen. 10), Transjordanian stories of Abraham (Gen. 14), Jacob (Gen. 31), and Esau (Gen. 36), and the stories of the Israelite campaigns against Sihon and Og in Transjordan in Numbers and Deuteronomy. For the Transjordanian place-names on the tribal map, Arias naturally used the tribal lists of half-Manasseh, Gad and Reuben in the book of Joshua, together with names of Edomite families in Genesis 36. Like Cranach, Ziegler, Wissenburg, Mercator, and Tilemann Stella before him, he included also the names of the halting sites (*mansiones*) of the Israelites on their journey from Egypt to Canaan from Num. 33 (though he finds 43 names in place of the usual 42). Arias's maps are, as one might expect, strictly the work of a biblical scholar, who reveals his biblical interests especially in the third map. His maps are neither particularly modern nor particularly original for their date, and do not make any new contribution to the development of the mapping of Transjordan.

CHRISTIAN ADRICHOM (ADRICHOMIUS)

Christian Adrichom (Adrichomius) was born in Delft on 14 February 1533 and died in Köln, where he had sought refuge from religious pressure, on 20 June 1585, aged 52. His *Theatrum Terrae Sanctae et biblicarum historiarum cum tabulis geographicis aere expressis* was published after his death, in 1593, at Köln by J. Birckmann. It contains a complete map entitled *Situs Terrae Promissionis SS Bibliorum intelligentiam experte aperiens: per Christianum Adrichomium Delphum*, together with separate maps for each tribe, and a map of Jerusalem. Transjordan is covered by maps of the tribes *Dimidia tribus Manasse*, *Ruben*, and *Gad*. There is an index of place-names for each tribe. Adrichomius also gives a very full *Index auctorum quibus usi sumus*, which reveals that he used the Latin translation of Santes Pagninus, and a full range of ancient and more recent scholarship including Burchard, Ziegler, Mercator and Laicksteen. His Preface also speaks of his sources and his intentions:

> Idcirco ad Dei T. Opt. Maximi gloriam & clariorem S. Scripturae intelligentiam, ex intimis eiusdem scripturae visceribus & medullis, ex probatissimis quibusque Auctoribus, ex Iosepho, Hieronymo, Egesippo, Plinio, Ptolomeo, Strabone, Brocardo, Breitenbachio, Saligniaco & multis aliis . . . maxima cum industria ac diuturna diligentiaque triginta annorum labore geographice delineavi, observata ubique iusta, quoad fieri potuit, locorum symmetria & commensuratione.

> Therefore to the glory of the highest and greatest Triune God and the clearer understanding of the Holy Scriptures, from the inmost organs and veins of the same Scripture, from all the most tried and tested authors, from Josephus, Jerome, Hegesippus, Pliny, Ptolemy, Strabo, Brocardus, Breitenbach, Saligniacus and many others . . . with the greatest industry and the daily diligence and labour of thirty years I have sketched this out in geographical manner, with the measurements and relationships of places observed as accurately as possible.

Adrichomius thus worked for over thirty years to the glory of God and the clearer understanding of Sacred Scripture, observing the strictest possible accuracy in the location of places. He was clearly a scholar of wide reading and intelligence, and his map contains more place-names than does any of his predecessors. Of particular value is the fact that Adrichomius' *Index* gives source references for the place-names used. Checking the place-names from the Transjordanian tribes on his index against Transjordanian place-names shown on the maps of his predecessors, one sees immediately that of the Index's 221 Transjordanian place-names 96 are additional to those appearing on the maps of Ziegler, Mercator, Tilemann Stella, Laicksteen and Arias

Montanus. Adrichomius' source references allow us to pinpoint where he found this extra information. Some 50 of these extra names came from the Bible (almost entirely from the Old Testament and Apocrypha); thus Adrichomius drew names like Amana, Senir, and piscinae Esebon from Song of Songs (4.8, 7.4), and Oronaim, Avran (= Hauran), and Adrach, from the prophets (Isa. 15.5, Ezek. 47.16, Zech. 9.1); he quarried a group of names (Carnion, Coelesyria, Alimis, Barasa, Bascama, Casbon) from the books of the Maccabees, and a number of lesser known names from the Pentateuch and the historical books; e.g., Save (= Shaveh-kiriathaim, Gen. 14); Thophel (Dt. 1.1); Camon (Jud. 10.5), Rogelim (2 Reg. [= 2 Sam.] 17. 27); Abana and Pharphar (4 Reg. [= 2 Kings] 5.12). He invents cities; thus Nabatha is given as a city founded by the Nebaioth (Gen. 25.13), and Edom and Galaad are given each their own *urbs*. Several places (Arabia, Bethsaida, Bosor, Deblatha, Iasa, Moab) are given their own *desertum*. Basan and Trachonitis are each given their own *via*. Adrichomius produced sites for Moses' view over the promised land, for the braying of Balaam's ass, for the Israelites' battle against Sihon king of the Amorites, for Paul's conversion, and even for the place where Jesus slept on board ship in the Sea of Galilee. Other examples could be given; clearly, Adrichomius makes the most of his biblical sources. He also adds 10 new names from Josephus, one item from Strabo (an unnamed lake near Gadara), 17 new names from Jerome, and 15 new names from William of Tyre. Adrichomius probably wanted to meet the needs of readers interested in the Crusades as well as the biblical stories.

The outlines of Adrichomius' map, like those of Arias Montanus two decades earlier, go back in essence to the 'modern' maps of the Ptolemaic atlas and ultimately to the map of Sanutus. The map is laid out with east at the top and the Palestinian coast line at the bottom. However, the coastal and river outlines do not improve on those of Ziegler and Mercator. The Palestinian coast shows no curve westwards towards the Nile Delta and no great sense of the topography of the coastline from Mt Carmel to

FIGURE 21. C. von Adrichom, 'Situs Terrae Promissionis SS Bibliorum intelligentiam exacte aperiens: per Chr. Adrichom' (from Mercator, *Atlas*, Eng. trans., II [1641]). (*The Board of Trinity College, Dublin*)

Ptolemais. The Kishon flows directly from the Sea of Galilee to the Mediterranean. The Jordan has lost the deep bend south of the Sea of Galilee seen on earlier maps, and in general has straightened out a little between the Sea of Galilee and the Dead Sea. However, Adrichomius shows little genuine knowledge of the Transjordanian tributaries (Fig. 21). From the east a large Hieromace fl[umen] pours into the middle of the eastern side of the Sea of Galilee. The Iaboc enters the Sea of Galilee at its southern end, close to the exit of the Jordan southwards. An unnamed river enters the mid-Jordan from the east, approximately where the modern Jabbok enters. One smaller stream enters the Jordan from the Dibon region. East of the Jordan, the Arnon torrens, rising near Philadelphia, flows south parallel with the Jordan (as in Ziegler and other maps) before turning west to enter the Dead Sea on its north-east shores. The Zareth torrens enters the Dead Sea half-way down its eastern shore. It is as if these eastern rivers have all been displaced northwards relative to the Sea of Galilee-River Jordan-Dead Sea, or as if Adrichomius has confused the Jabbok with the Jarmuk and the Zered with the Arnon (leaving him with no identification for the river entering the Jordan half way between the Dead Sea and the Sea of Galilee). The Dead Sea has a curious twist to the west at its southern, pointed end. The Red Sea (Mare Rubrum sive Sinus Arabicus) has one arm only, with no indication of its division into two gulfs, one either side of the Sinai peninsula. Geographically and scientifically, Adrichomius' map was outdated from the start. Unfortunately it was accepted as the major reliable source for Holy Land cartography for at least the next century. It must be said that the maps of Arias Montanus and Adrichomius, for all the biblical information incorporated, added little to the serious development of the cartography of the Transjordanian region.

The sixteenth century throughout shows the influence of Ziegler and Mercator. It shows also the continuing importance of the theme of the Exodus and the wilderness wanderings of the Israelites (cf. Mercator, Wissenberg, Münster, Tilemann Stella [1557], Laicksteen, Arias Montanus). Maps continued to show the boundaries of the Israelite tribes, and Stella emphasised on his first map the kingdoms of Judah and Israel. The Bible, in short, continued to provide both the source and the motivation for the sixteenth-century maps of the Holy Land. In the second half of the century, scholars like Arias Montanus and Adrichomius put their formidable biblical scholarship to the service of cartography, and produced maps showing more biblical sites than ever before. But there was little sign in these maps — except for the work of Laicksteen and Sgrooten — of any serious first-hand exploration of the region, and no sign of interest in the landscape, towns and villages of contemporary Palestine. In this there was a marked regression from the period of the Crusades.

However, cartographical techniques were developing. Though Münster's maps were printed from woodcuts and were not unlike Ziegler's, Mercator and his successors used engraved copper plates to produce individual sheet maps or wall-charts of six or nine sheets combined. Sixteenth-century cartographers approached the Holy Land from different angles; thus Mercator presented a view of the Holy Land from the south-east, looking north-west, with the Mediterranean coast concave across the top of the map, in which he was followed by Stella's first map. Later maps showed the Mediterranean coast running diagonally across the map from top right to lower left, while the Laicksteen-Sgrooten map drew the Mediterranean coast running vertically from top left to bottom left of the map. Arias Montanus and Adrichomius, however, showed their conservatism by reverting to the earlier presentation of the Holy Land with east at the top and the Mediterranean coast at the bottom.

TABLE 5. Transjordanian toponyms from Ziegler (1536) to Mercator (1606)

Ziegler 1536	*Mercator 1537*	*Wissenburg 1538*	*Munster 1540*	*Stella 1552*	*Stella 1557*	*Laicksteen & Sgrooten 1570*	*Laicksteen & Sgrooten inset 1570*	*Mercator 1606*
	Abarim m.	Abarim montes	Abarim mos	Habarim mons	Abarim montes	Mons Abarim		Abarim mons
Abila	Abel que e vineis			Abila Vinearum			Abela	
Abila Lysanii								Abila sisanii
	Achie, Achalgai			Achie et Achalgai	Achie			
								Acrabatene toparchia
						Adaim	Adaim	Adai
	Adam					Adama		Adama
	Edom		Aedom [Edom]					
						Alexandria		Alexandria
Halmon				Almon				
Diblathaimah				Diblathaim				
						Ammaus		Ammaus
						Ammaus toparchia		Ammaus toparchia
Amathus	Amathus			Amathus				
Ammonitis		Ammonitae	Ammonitae	Ammonitis regio quae nunc Philadelphia	Ammonitarum regio vel Philadelphia		Ammonitarum	
Amoreus mons	Amorrhaeus mons		Amorrhaei	Amorrhaei		Amorrhaei		Amorrhaei
				Antilibanus mons	Antilibanus mons			
Arabiae desertae pars					Arabiae desertae pars			
					Arabia Nabathea			
Petraea Arabiae desertae pars	Arabia Petraea		Arabia Petraea	Arabia Petraeae pars	Arabia Petrea		Arabiae Petraeae pars	
						Araboth		Araboth
						Araboth		Araboth
						Arama		Arama
						Arba		Arbel
	Arta			Arta				
						Areopolis castrum		Areopolis castrum
Argob	Argob urbs regio		Argob	Argob regio	Argob	Argob		Argob
Arnon	Arnon torrens	Arnon torrens	Arnon fl.	Arnon flu.	Arnon flu.	Arnon torrens		Arnon torr.
						Arnon schiz.		Arnon schiz.

Ziegler 1536	*Mercator 1537*	*Wissenburg 1538*	*Munster 1540*	*Stella 1552*	*Stella 1557*	*Laicksteen & Sgrooten 1570*	*Laicksteen & Sgrooten inset 1570*	*Mercator 1606*
Haroher	Aroer	Aroer	Aroer	Aroer	Aroer	Aroer	Aroer	Aroen
		Aroer (by Rabbath)						
	Asedoth phasga			Asdoth pisga	Asdoth			
	Astaroth (x 2)		Astorot	Astaroth	Astarot	Asteroth		Asteroth
								Astaroth (2)
						Ataroth		Ataroth
						Athona		Athona
						Baalag		Baalag
						Baalgad	Baalgad	Baalgat
Bamoth vallis	Bamoth vallis			Bamoth vallis		Bamoth vallis	Bamoth vallis	Bamoth vallis
						Barasa		Bara
	Basan regnum		Regnum Basan	Basan reg.		Regnum Basan		BASAN REGNUM
Bathenea	Bathanea		Bathanea	Batharea		Bathenaei		Bathenaei
Beasthra	Beasthra			Beasthra				
Bosor	Bosor			Bosor	Bosor			Bazor desert/
	Baalmeon			Bethbaalmeon		Beelmeon		Beelmeon
Puteus	Beer Puteus		Beer Puteus	Beer Puteus			Beer puteus	
		Bethabara		Bethabara desertum		Bethabara		Bethabara
Betharan	Betharam			Betharam				
Bethhaiesimaoth	Bethjesimoth		Betiessmoth	Bethhaiesimoth		Bethiessimoth		Bethiesimoth
	Betonim			Betonim		Betonim		Bethonim
Bethnimrah	Bethnemra			Bethnimrah				
								Bethpeor
						Bethsabara		Bethsabara
						Bethsemes		Bethsemes
Caesarea	Cesarea Philippi	Caesarea Philippi	Cesarea Philippi	Caesarea Philippi	Cesarea Philippi	Caesarea Philippi	Caesarea Philippi	Caesarea Philippi nunc Bellena
					Catachrysea			
Callirhoe	Callirhoe					Calliroe		Callirrhoe
	Campestria Moab	Campestria Moab		Campestria Moab		Campestria Moab		
Canatha	Canatha			Canatha				
Capitoliada	Capitoliada			Capitolias	Capitolias			
	Charchar			Charchar		Carchar		Carchar
						Carioth		
						Carnaim		Carnay
						Casphor		Casphor
						Cedar		Cedar

Ziegler 1536	*Mercator 1537*	*Wissenburg 1538*	*Munster 1540*	*Stella 1552*	*Stella 1557*	*Laicksteen & Sgrooten 1570*	*Laicksteen & Sgrooten inset 1570*	*Mercator 1606*
	Cariathaim			Chiriathaim				
Coele Sayria	Coele Syria				Coele Syriae pars			
Colchiel	Colchiel				Colchiel			
								Copar
Chorozaim	Chorazim		Corocaim	Chorasim	Chorazaim	Corozaim	Corozaim	Corozaim
Damascus		Damascus			Damascus	Damascus	Damascus	Damascus
Dan	Dan		Dan	Dan	Dan	Fons Dan	Dan fons	Dan fons
	Damnaba			Damnaba				
						Daria		Daria
						Datheman, torrens		Datheman
Debir	Debir	Debir	Debir	Debir		Debir		Debir
						Dibon torrens		Debon
								torrens
Dedan			Dedan		Dedan, Dodanim		Dedan	
						Diblathaim		Diblathaim
	Dibon			Dibon		Dibon	Dibon	Dibon
Dibon Ghad	Dibon Gad	Dibongad	Dibon Gad	Dibon Gad				
						Dion		Dion
Disadab					Disabad			
						Ecbatana		Ecbatana
	Edom		Aedom					
Edrehi	Edrai			Edrei			Edraim	
		Elaniticus sinus						
		Elath						
	Eleale					Eleale		Eleale
						Elon		Elon
						Elon torrens		
						Emathitae		Emathitae
Ephron	Ephron			Ephron	Ephron	Ephraim	Ephron	Ephraim
						Esthomonitim		Esthomonitim
				Ferreus mons				
						Fons Arnon		
						Fons Carnaim		
Ghad	Gad	Gad	Gad	Gad				Gad
						Gadera seu Gadara		Gadara
				Gadarenus lacus				
						Gadera seu Gadara		Gadera
	Galaad		Galaad			Galaad aut		Galaad
						Masphath		
Gamala	Gamala			Gamala		Gamela		Gamela
Gaulon	Gaulon			Gaulon	Gaulon	Gaulana, Gaulanitae		Gaulana,
				Gaulanitis regio				Gaulanitae
					Gebalene regio			

Ziegler 1536	*Mercator 1537*	*Wissenburg 1538*	*Munster 1540*	*Stella 1552*	*Stella 1557*	*Laicksteen & Sgrooten 1570*	*Laicksteen & Sgrooten inset 1570*	*Mercator 1606*
Genesar	Genesar	Genezar lacus	Genesar		Genesare			
						Gennabara		Gennabar
		Gerasa		Gerasa		Regio Geraserorum	Geraza	Gerasaeorum
								regio
						Geraim		Geray
	Gergesa dicta			Gergesa				
	Geraa &							
	Gadara							
Gessuri	Gessuri		Gessuri	Gessuri vel				
				Batharea				
				Gilead mons				
				Gileaditis				
		Gomorra				Gomorra		Gomorra
Hermon	Hermon mons		Mons Hermon			Hermon mons	Mons Hermon	
						Hermon torrens		Hermon
								torrens
Chesbon	Hesebon	Esbon	Hesebon	Chesbon		Heshbon	Heshbon	Hesbon
					Hippus mons			
Iabis	Iabes		Iabes	Iabis	Iabis			Iabis
						Iabek	Iabok	Iaboc
Iabboc	Iaboc torrens	Iaboc torrens	Iaboc fl.	Iaboc flu.	Iaboc flu.	Torrens Iaboc/Iabec		Iaboc torrens
	Iaboc flu.							(of two diff.
								rivers)
Iogbehah	Iegbaa			Iogbehah		Iagabea		Iagabea
Iahasar	Iasa	Ianatz		Iahasah				
					Iathkeel			
Iahazer	Iazer				Iaazer	Iazer	Iazer	Iazer
	Idumaea	Idumaea	Idumaea		Idumaea	Idumaea inferior		
						Illustris vallis		Illustris vallis
Ior	Ior		Ior	Ior	Iar	Fons Ior		Ior fons
Iordanis			Iordanis fl.		Iordanis flu.			Iordanis
								fluvius
	Iturea	Iturea		Ituraea vel			Iturea seu	ITURAEA
				Chamat			Traconitides	
	Iulias ac			Iulias	Iulias	Iuliada		Iuliada
	Bethamarta							
Cademoth	Cademoth		Cademoth	Cademoth		Kademoth		Kademoth
solitudo	solitudo		solitudo	solitudo				
				Kibaleh				
			Krac					
				Lahasah				
Lasa	Lasa			Lasa	Lasa			
Libanus mons	Libani montis	Libanus mons	Libanus			Mons Libanus		Libanus mons
	pars							

Ziegler 1536	*Mercator 1537*	*Wissenburg 1538*	*Munster 1540*	*Stella 1552*	*Stella 1557*	*Laicksteen & Sgrooten 1570*	*Laicksteen & Sgrooten inset 1570*	*Mercator 1606*
Libyas	Libyas			Libias		Livias		Linias
Luchit				Luchit	Luchit			
				Maachati	Maachati			
						Macheronta		Macherunta
	Madian				Madian			
Magedon	Magedon			Magedon	Maghedon	Mageth		Mageth
Machanaim	Mahanaim	Mahanaim		Machanaim		Mahanaim		Mahanaim
Machane	Machane			Machane				
Machrus	Macherus			Macherus				
Manasse	Manasse		Manasse	Manasses	Manasse			Manasse
	Mare Galilea Cenereth & Tiberiadis. Apud Lucam Stagnum Genesareth		Mare Galileac			Mare Galilaeae aut Cinereth vel Tiberiadis	Mare Galilaeae	Mare Galilaeae vel Tiberiadis quod et stagnum Genezareth
Asphaltis	Mare mortuum	Mare mortuum seu salsum	Mare mortuum	Mare salsum vel mortuum	Lacus Asphaltites vel Mare mortuum aut salsum	Lacus Asphaltites vel mare mortuum aut salsum	Mare mortuum	Mare mortuum, salsum, sive lacus asphaltitis olim vallis salinarum
					Luchit			
	Masphe					Masspha		Maspha
						Masphath	Maspath	Masphath
Mathana solitudo	Mathana solitudo		Mathanae solitudo	Mathanah solitudo				
Medba	Medaba	Medba		Medba		Medba		Medba
Medebar	Medebar			Medebar				
	Mephahat			Mephaath		Mephaath		Mephaath
Minnith	Minith			Minnith				
				Mispeh				
Moab			Moab			Moab	Moab	Moab desertum
	Moabitis regio	Moabitarum regio		Moabitis regio	Moabitarum regio		Moabitis regio	Moabitarum regio
						MonsArnon		Mo Arnon
						Mons Bethoron		Mons Bethoron
	Galaad mons					Mons Galaad	Mons Galaad	Mons Galaad
	Nebo mons			Nebo mons				Mons Nebo
								Mons Pisga

Ziegler 1536	*Mercator 1537*	*Wissenburg 1538*	*Munster 1540*	*Stella 1552*	*Stella 1557*	*Laicksteen & Sgrooten 1570*	*Laicksteen & Sgrooten inset 1570*	*Mercator 1606*
						Mons Sener		
		Montana Felicis Arabiae						
						Naballa		Naballa
				Nabathaei				
Nachaliel	Nahaliel		Nahaliel	Nachaliel	Nathahel	Nahalihel	Nahalihel	
						Nain		Nain
						Nemra		Nemra
					Neue			
Nobach	Nobe			Nobach				
						Pannadaei		
						Panneas		
Pella	Pella		Pella	Pella	Pella	Pella		Pella
	Peraea regio	Peraea		Peroea regio		Peraea		PERAEA
Petra regia	Petra deserti	Petra regia	Petra deserti		Petra, Sela, Arachen, Recem, Iaktheel, Nabath		Petra Crach	
Phiala	Phiala fons			Phiala fons		Fons Phiala		Phiala fons
								PHILADELPHIA
Philadelphia	Rabbath apud Grecos Philadelphia	Rabbath Ammon	Rabbath	Philadelphia	Philadelphia	Philadelphia	Rabbath	Philadelphia
						Rabbath		Rabbaath
Punon	Phunon	Phunon	Phunon				Phunon	
Penuel	Phanuel			Penuel	Peruel	Pniel		Pinel [ie Peniel]
								Pisga torrens
	Ramoth		Ramoth			Rama Galaad	Rama Galaad	Rama
				Ramath			Ramathi	
		Rhabmathon						
Ruben	Ruben	Ruben	Ruben	Ruben	Ruben			Ruben
						Rydda		
						Sabama		Sabama
						Saharum		Sabarim
						Salabim		Salabim
Salcha				Salcha	Salcha			
Samachonitis	Samachonitis			Samachonites lacus	Samachonites lacus	Lacus Samachonit-is vel maron	Lacus Samachonitis	Samachonitis lacus
Samuli	Samuli			Samuli	Samuli			
Saphon	Saphon			Saphon				
								Scham, olim Damascus
						Sebeth		Sebeth
								Seboim

Ziegler 1536	*Mercator 1537*	*Wissenburg 1538*	*Munster 1540*	*Stella 1552*	*Stella 1557*	*Laicksteen & Sgrooten 1570*	*Laicksteen & Sgrooten inset 1570*	*Mercator 1606*
	Segor		Segor					
					Sela			
Seleucia	Seleucia		Seleucia	Seleucia		Seleucia		Seleucia
						Sepulchrum Iob		
	Seir mons	Seir mons	Seir montes			Seyr mons aut Cedar		Seir mons (SW of Dead Sea)
						Sittim	Sittim	Sittim
	Sodoma	Sodom	Sodoma			Sodoma		Sodama
						Sophan ataroth		Sophan
	Sochot					Succoth	Succoth	Succoth
						Tarahassa		
	Telemanum			Telemannum	Telemannum			
						Terra Epher		
	Terra Hus		Terra Hus			Terra Hus aut Traconitidis/Terra Hus seu Traconitis		Terra hus
						Terra Macces		Terra macces
	Theman		Theman		Theman	Theman (Iturea)	Theman (Iturea)	Theman (Iturea)
						Thermae		Therme
Traconi	Thraconi montes			Thraconi montes				
						Tibebath		Tibebath
						Torrens Seyr		
	Thraconitis	Traconitis	Traconitis	Traconitis regio		Terra. . . Traconitidis/ Traconitis		Trachonitis regio
						Tubbin		Tubbin
						Turris Libani		Turris Libani
						Vallis Hermon		Vallis Hermon
	Zeb		Zeb	Zeb	Zeb			
Zered torrens	Zared		Zared torrens	Zered torrens	Zered flu.		Zared torrens	
						Zoba		Zoba
Sueta	Sueta		Sueta	Sueta	Sueta	Swetha	Zwetha	Zwetha
+ mansiones	+ mansiones 1-41	+ mansiones 1-41	& mansiones 32-41		+ mansiones 1-41		+ mansiones 1-42	

CHAPTER SIX

SEVENTEENTH-CENTURY PUBLICATIONS

JODOCUS HONDIUS (1563–1612) AND CORNELIS CLAESZ (1560–1609) AND SUCCESSORS

Mercator's original *Atlas* was published in 1595 and again in 1602. This first edition contained no map of Palestine. In 1606 a new version of the Atlas was published by Jodocus Hondius and Cornelis Claesz. This included a map (Fig. 22) entitled 'TERRA SANCTA/ quae in Sacris/Terra Promissionis ol./PALESTINA' (Koeman 1969, 8150:1A), which drew on the Laicksteen-Sgrooten map (though re-orienting it with east at the top). This map has ignored Mercator' map of 1537, and has followed the outline provided by the Laicksteen-Sgrooten map, with its distinctive version of the Sea of Galilee, the Dead Sea, and the course of the Jordan linking them (though some changes have been made in the identification of the rivers: the Datheman Torrens, running into the south-east shore of the Sea of Galilee, has become the Iaboc Torrens, and the Arnon Torrens appears only briefly in the east, making no connection with the Dibon Torrens as in Laicksteen's map). The accompanying text in the 1606 *Atlas*, however, makes clear that both the Jordan and the Arnon enter the Dead Sea: the Jordan 'post illustrem convallem praeterlapsus, post multos ac sinuosos flexus iuxta Arnonis ostium velut invitus in Mare Mortuum sive Lacum Asphaltitem ... dirum natura illabitur' (p. 326) ('having flowed past the famous valley' [i.e., the Esdraelon], 'past many winding bends, near the mouth of the Arnon falls as if unwillingly into the Dead Sea or Lake Asphaltites'). This map has also ignored the Transjordanian place-names Mercator used in 1537, and taken over the place-names of the Laicksteen-Sgrooten map. Strangely, this 1606 map notes the tribal territories for Cisjordan, but not for the Transjordanian region, and the names of Ruben, Gad and the Transjordanian half of Manasseh do not appear.

This Mercator-Hondius *Atlas* ran into a number of editions, in Latin and in French, under the names of Jodocus Hondius, Jodocus Hondius II (1594–1629), and Henricus Hondius (1597–1651) (see Koeman, revised, 1 [1997], 5 for the full list). This Laicksteen–Sgrooten approach made a re-appearance (see below) in W. J. Blaeu's map of 1629 (Blaeu, G., *Appendix Theatri A. Ortelii et Atlantis G. Mercatoris*, 1631; Blaeu, J., *Asia quae est Geographia Blaviana parts Quarta*, libri duo, volumen decimum, Amst., 1662, pp. 29–30, and in C. J. Visscher's map of 1642 (*Geographische Beschryvinge van T'Beloofde Landt Canaan*, Amst. 1642; cf. Nebenzahl 1986, pl. 46, 124–25), but thereafter it lost favour and yielded to the growing popularity of Adrichom's work.

In 1630 and 1631 Johannes Janssonius and Henricus Hondius respectively published an *Appendix*. These Appendices contained a map entitled 'PALAESTINA, SIVE TERRAE SANCTAE DESCRIPTIO' (Koeman 8150:1B), bordered top and bottom by nine scenes from the Exodus story, with a small inset map-view of Jerusalem. This map drew on Adrichom's map, though it makes the Mediterranean coast of Palestine bend westwards towards the Nile Delta (cf. Fig. 31) instead of continuing in a straight line. East of the Jordan valley, it remains close to Adrichom's presentation (for example, it still shows the regio Edom and Seir mons east of the Sea of Galilee), but omits the names Hieromace and Jaboc from the two rivers running westwards into the Sea of Galilee, and also a number of minor place-names. (This map was later reproduced in J. Jansson's *Atlas Contractus*, vol. 2, 1666, and by G. Hornius in his *Accuratissima Orbis Antiqui Delineatio*, several editions, 1652–84.)

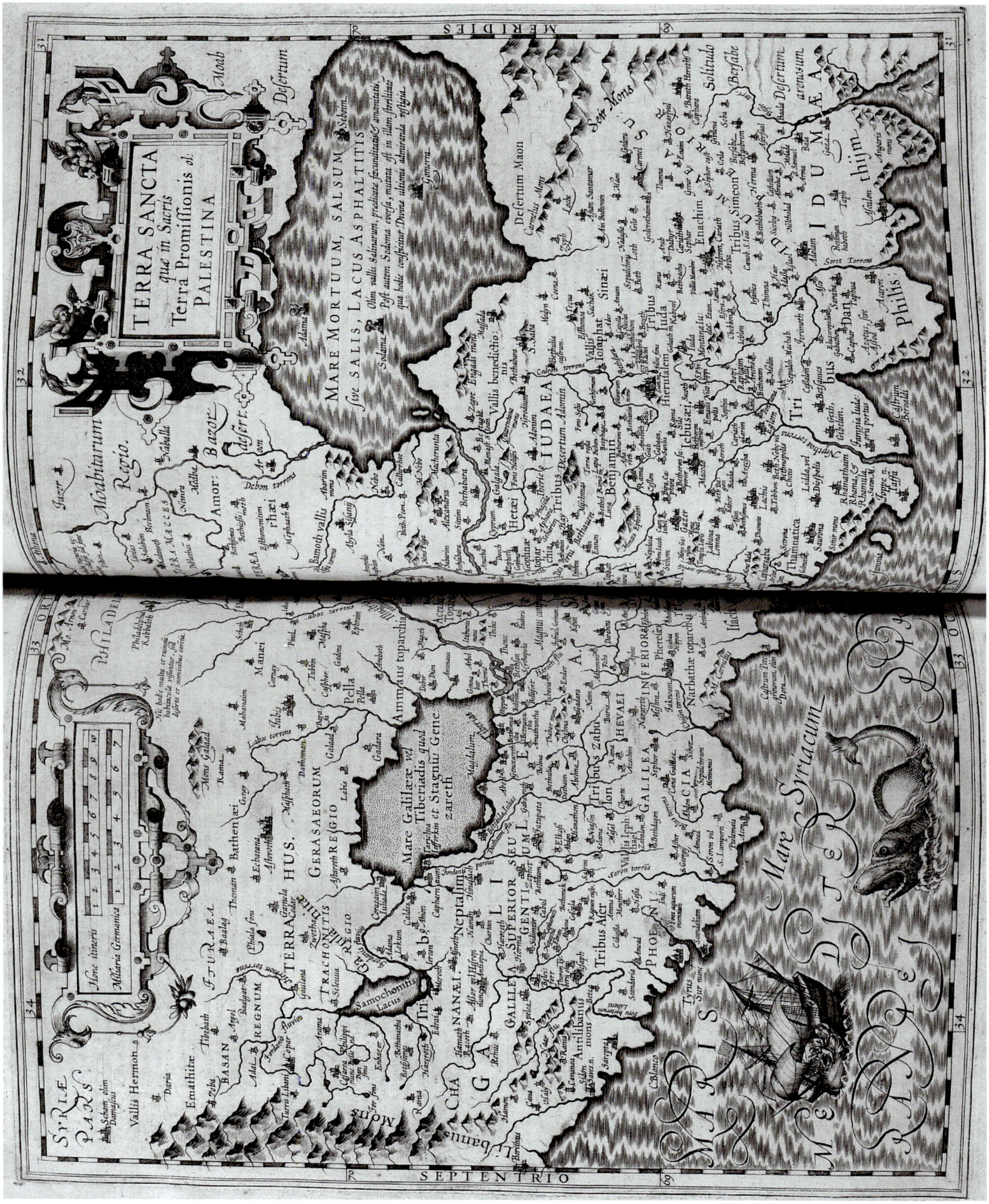

FIGURE 22. G. Mercator, 'Terra Sancta quae in Sacris Terra Promissionis ol: Palestina' (*Atlas 1606*). (*The Board of Trinity College, Dublin*)

JOHANNES JANSSONIUS (1588–1664), HENRICUS HONDIUS (1597–1651), AND GEORGE HORNIUS(1620–70)

The Mercator-Hondius *Atlas* was taken over by Johannes Janssonius, and various editions in French, German, Dutch and English were published in the 1630s (see Koeman, revised, I [1997], 5–6 for list). These editions presented the Adrichom map entitled 'SITUS TERRAE PROMISSIONIS S.S. Bibliorum intelligentiam exacte aperiens/per Chr. Adrichom' (Koeman 8150:1C.1).

In 1638 a new three-volume version of the *Atlas* appeared under the names of Johannes Janssonius and Henricus Hondius, the *Atlas Novus*. It subsequently appeared in four volumes (1647), five volumes (1649–59), and in six volumes (1657–62), under the sole name of Johannes Janssonius. The Adrichom map of Palestine (Koeman 8150:1C.1) appeared in the 1638 edition, and also in the six-volume edition *c.* 1662 (Koeman 8150:1C.2; see Koeman, revised, I [1997], 705). The climax of this publishing sequence was the *Atlas Major* published by the Janssonius heirs in 1675, in which the map 'PALAESTINA, SIVE TERRAE SANCTAE DESCRIPTIO' (a revised version of Adrichom's map; Koeman 8150:1B; Koeman, revised, I [1997], 705) reappeared. (This whole sequence of atlases, culminating in the *Atlas Major*, in addition to the original Latin edition, appeared also in French, German, and Dutch editions; the four-volume edition appeared also in Spanish and English versions [for lists see Koeman, revised, I (1997, 7–9].)

In 1666 the Janssonius heirs published the *Atlas Contractus* in two volumes (*Johannis Janssonius Atlantis Contracti Pars Posterior*, Amst. Apud Joannis Janssonii haeredes, Anno MDCLXVI). This contained the revised version of Adrichom's map entitled 'PALAESTINA, SIVE TERRAE SANCTAE DESCRIPTIO' map (Koeman 8150:1B) which had previously appeared in the *Appendix* maps of 1630 and 1631.

In 1652–57 Johannes Janssonius published four editions of Georgius Hornius' *Accuratissima Orbis Antiqui Delineatio*, followed by *Accuratissima Orbis Delineatio* in 1660; this title was republished by Johannes Janssonius van Waesbergen in 1677 and 1684. These atlases (Koeman 1:601/7, listed in Koeman, revised, I [1997], 10) contain a map of the Holy Land divided into six sheets (Koeman 8150:1D/1–6; Koeman, revised, I [1997], 705–6), with the following titles: (1) TRIBUS RUBEN ... (2) Tribuum EPHRAIM ... (3) TRIBUS ASER ... (4) DIMIDIA TRIBUS MANASSE ... (5) Pars maxima TRIBUS IUDA ... (6) TRIBUS SIMEON .... (These maps also appear in the six-volume edition of the *Atlas Novus*, in the Latin, French and Dutch versions, and in the textless five-volume edition of 1680.) This set of six sheets is based on, and drew its place names from, the Adrichom map.

These Hornius atlases contained two further maps of the Mercator tradition. The first is a map entitled 'IUDAEAE/seu/TERRAE ISR/AELIS/Tabula geographica;/in qua/locorum in Veteri et No/vo Testamento celebratissimo/rum Situs accurate descrip/ti/ arte facta/ a Tilemanno Stella Sigenensi' (Koeman 8155H:1; Koeman, revised, I, 706). This map (Fig. 23) is derived directly from Stella's 'Typus chorographicus celebrium locorum in regno Iudae et Israhel' published in Ortelius' *Parergon* (1586–1624). The second is a map entitled 'TABULA ITINERARIA/ PATRIARCHARUM/ABRAHAMI, ISAACI/et/JACOBI' (Koeman 8020H:1; Koeman, revised, I [1997], 702). This map is based on the map originally published by Ortelius in 1586 and subsequently in several editions of *Theatrum Orbis Terrarum* entitled 'ABRAHAMI PATRIAR-/CHAE PEREGRINATIO,/ ET VITA' (Koeman 8020J:31A and 31B; Koeman, revised, IIIB [2003], 866). The place-names are taken from the patriarchal stories of the book Genesis, especially Gen. 14, but this later version omits reference to the people Zuzim from Gen. 14.5, and adds the place Pniel from Gen. 32.30 ('Pniel est trajectus ad amnem Iabbok').

WILLEM JANSZOON BLAEU (1571–1638)

The Laicksteen-Sgrooten branch of the Mercator-Hondius tradition, however, continued into the seventeenth century through the work of the Blaeu family. (On the Mercator-Hondius tradition, see Keuning 1947, 37–62; 1948, 63–71.) In Mercator's *Atlas* (Amsterdam 1606 and 1620) Jodocus Hondius I published a map entitled 'Terra Sancta quae in Sacris Terra Promissionis olim Palestina'. A new version of this tradition, which drew its inspiration

FIGURE 23. A derivative of Stella's 1552 map, published in G. Hornius' *Accuratissima Orbis Delineatio* (1677). (*The Board of Trinity College, Dublin*)

from the Laicksteen-Sgrooten inset map of 1570, was published under the same title by Jodocus Hondius II, bearing the date 1629. Jodocus died the same year, aged 34, and his work was taken over and published immediately by Willem Janszoon Blaeu, the name of Gulielmus Blaeuw being substituted immediately on the cartouche for that of Jodocus Hondius (Blaeu, *Appendix Theatrii A. Ortelii et Atlantis G. Mercatoris*, Amsterdam 1631, and later *Asia quae est Geographiae Blavianae pars Quarta*, libri duo, volumen decimum, Amstelaedami, J. Blaeu, 1662; see now Blaeu 2005, 502–03). This map is distinctive in several ways. It presents the Mediterranean coastline along the top, the Egyptian coast angling off towards the top left corner. The Dead Sea follows the Laicksteen-Sgrooten model, with a sharp eastward turn at the southern end. Three main rivers enter the Jordan and Dead Sea from the east: the Jabbock enters the Jordan between Mahanaim and Pniel, nearly half-way between the Sea of Galilee and the Dead Sea (thus no longer close to the southern end of the Sea of Galilee); the Arnon enters the Dead Sea at its northeastern, and the Zared near its south-eastern corner. (Unfortunately, in Cisjordan Blaeu continued the ancient error of linking the Sea of Galilee and the Mediterranean directly by the river Kishon.) The modern Gulfs of Suez and 'Aqaba are clearly divided by the Sinaitic peninsula. Comparison of the lists of Transjordanian place-names included in various maps, and the presence of the names Camon, Elath, Evila desertum, Holen, Hippon, Rabba, Resenorum [= Gerasenorum?] Terra, Zartan, Terra Tob, Tesbe, and Vallis Zaue, suggest that the immediate source of the hundred or so names used was Adrichom (1590), though a number of names go back to the earlier maps of Mercator and Laicksteen.

NICOLAS SANSON (1600–67)

Important also were the maps of Gullaume (d. 1603) and Nicolas Sanson (1600–67) of Abbeville. Gullaume, the father of Nicolas, produced a map 'Judaea seu Terra Sancta quae Hebraorum sive Israelitarum in suas duodecim Tribus divisa ab invicem Regnis Juda et Israel expressis . . . ex conatibus geographicis Gulielmi Sanson', which was published in N. Sanson's *Atlas Nouveau contenant toutes les parties du monde* (Paris: H. Jaillot, 1696). In the same volume Sanson published an interesting map 'Les Deserts de l'Egypte . . .' (no. 60), which looked down from the south-east on the Egyptian, Arabian and Syrian deserts, with the Palestinian coast and the Mediterranean in the upper centre of the map. In the late seventeenth century the publisher F. de Wit published an *Atlas* in Amsterdam: this contained a map entitled 'Terra Sancta sive Promissionis olim Palestina recens delineata et in lucem edita per Nicolaum Visscher, Anno 1659' (cf. Nebenzahl 1986, pl. 49, pp. 132–33). This map clearly follows the Adrichom outlines, but like Blaeu's map of 1629 and Mercator's original map of 1537 sets the Mediterranean coast at the top, not the bottom; the Transjordanian region appears in the foreground, but in the central foreground, east of the Solitudo Kademoth, appears a view of the Israelite encampment in the wilderness. Nicolas Sanson also produced a map 'Tabula Secunda in qua Terra Promissa in suas Tribus partesque divisa' in his *Geographia Sacra ex veteri et novo Testamento desumta* (Amsterdam, 1711). (Bound into the same volume is an edition by J. Clericus of Eusebius' *Onomasticon*, ed. by J. Bonfrerius, and a text of Brocardus' *Descriptio Terrae Sanctae*. Between these two works is a copy of Bonfrerius' map 'Tabula Geographica Terrae Sanctae', from the original in Bonfrerius' *Josue, Iudices et Ruth commentario illustrate* [Paris, 1631].)

The continental maps of the Holy Land and Transjordan in the seventeenth century, though colourful and popular with collectors, are cartographically disappointing. The influence of Adrichomius (which was regressive rather than progressive) is evident everywhere; in the publications of the Blaeu family, the Laicksteen-Sgrooten tradition was overlaid by the Adrichomian format, though the Mediterranean coast was now printed at the top instead of the bottom of the sheet. In short, there was no great scholarly development in the cartography of the Holy Land in the seventeenth century, at least among the great continental mapmaking houses.

CHAPTER SEVEN

JOHN SPEED AND THOMAS FULLER

JOHN SPEED (1552–1629)

JOHN SPEED, born the son of a tailor in Cheshire, became an historian and mapmaker, best known for his county maps of England. His first collection of maps was published under the title *The Theatre of the Empire of Great Britaine* in 1611: this was reprinted in 1614, with a Latin version in 1616, and again in 1623. Augmented by *A Prospect of the most famous Parts of the World* this reappeared in 1627; this was reprinted a number of times (1631–32, 1646, 1650–54, 1662–65), but in 1676 a new edition appeared, enlarged by E. Phillips, with new maps of Canaan, East Indies, Russia and other places. John Speed also produced a small book entitled *The Genealogies recorded in the Sacred Scriptures according to every Family and Tribe with the line of our Saviour Christ observed from Adam to the Blessed Virgin Marie*; this first appeared in 1611 and between then and 1640 was reprinted many times, often combined with Prayer Book and Psalter.

Speed's maps of Canaan make an important study in themselves. His first map was a wall map in four sheets entitled *Canaan as it was possessed in Abraham and Israel's dayes, with the stations and bordering nations*. A cartouche conveys a dedication and a date: 'To the reverent and worshipful maiester W. Cotton doctor of divinitye I.S. wisheth eternall blessedness in Chryste 1595.' His second map, *Canaan*, was originally dated 1611, with a cartouche bearing the statement 'Begun by Mr John More continued and finished by John Speed'; in the 1676 version of *The Theatre . . . together with A Prospect . . .* the date was manually altered to 1651. A smaller, quarto, version of this map was published in Speed's *Genealogies*. The date of 1611 coincided with the publication of the King James Bible and Speed's *Genealogies*.

Speed's 1595 map explicitly copies the maps of B. Arias Montanus, whose work is acknowledged in miniscule type below the title. The outlines of the land, its inland seas and rivers, follow closely those of Arias, and the place-names are taken, with some omissions and some changes of spelling, from Arias' two maps *Tabula terrae Canaan Abrahae tempore* and *Terra Israel omnis ante Canaan . . . tabula*. Speed follows Arias in locating the tribe of Simeon to the east of the Dead Sea.

Speed's 1611/51 map (first published in 1617, according to the catalogue *Undique ad Terram Sanctam: Cartographic Exhibition from the Eran Laor Collection*, 1976, p. 18) is quite different (Fig. 24), and approximates more to Ortelius' version of Tilemann Stella, though he presents a much more rugged coastline and a more complex river system. In the south, like Ziegler, Speed shows Ezion geber and Elath at the head of the eastern gulf of the Red Sea. Where his predecessors drew the Zered, entering the Dead Sea on its south-eastern shore, Speed drew the Dimon; he makes the Sered join the Arnon from the south-east and the Waters of Nimrim join the Arnon from the north before these three rivers, with sundry other minor streams, enter the Dead Sea at its north-east corner. The river entering the Jordan midway between the Dead Sea and the Sea of Galilee is not named; the Jabok enters the Jordan just south of the Sea of Galilee, roughly where the modern Wadi Yarmuk runs. Two sources of the Jordan are drawn in the usual fashion, but are not actually named as Jor and Dan. Speed correctly locates Corazin on the west coast of the Sea of Galilee. In this map, Adrichom rather than Montanus seems to have been the immediate source for most of the Transjordanian place-names; some names, such as Forest of Ephraim (Saltus Ephraim), Lodebar, Oronaim, Rogelim and Tishbe, do not appear in any other of Speed's possible sources. For some names (Admah, Betonim,

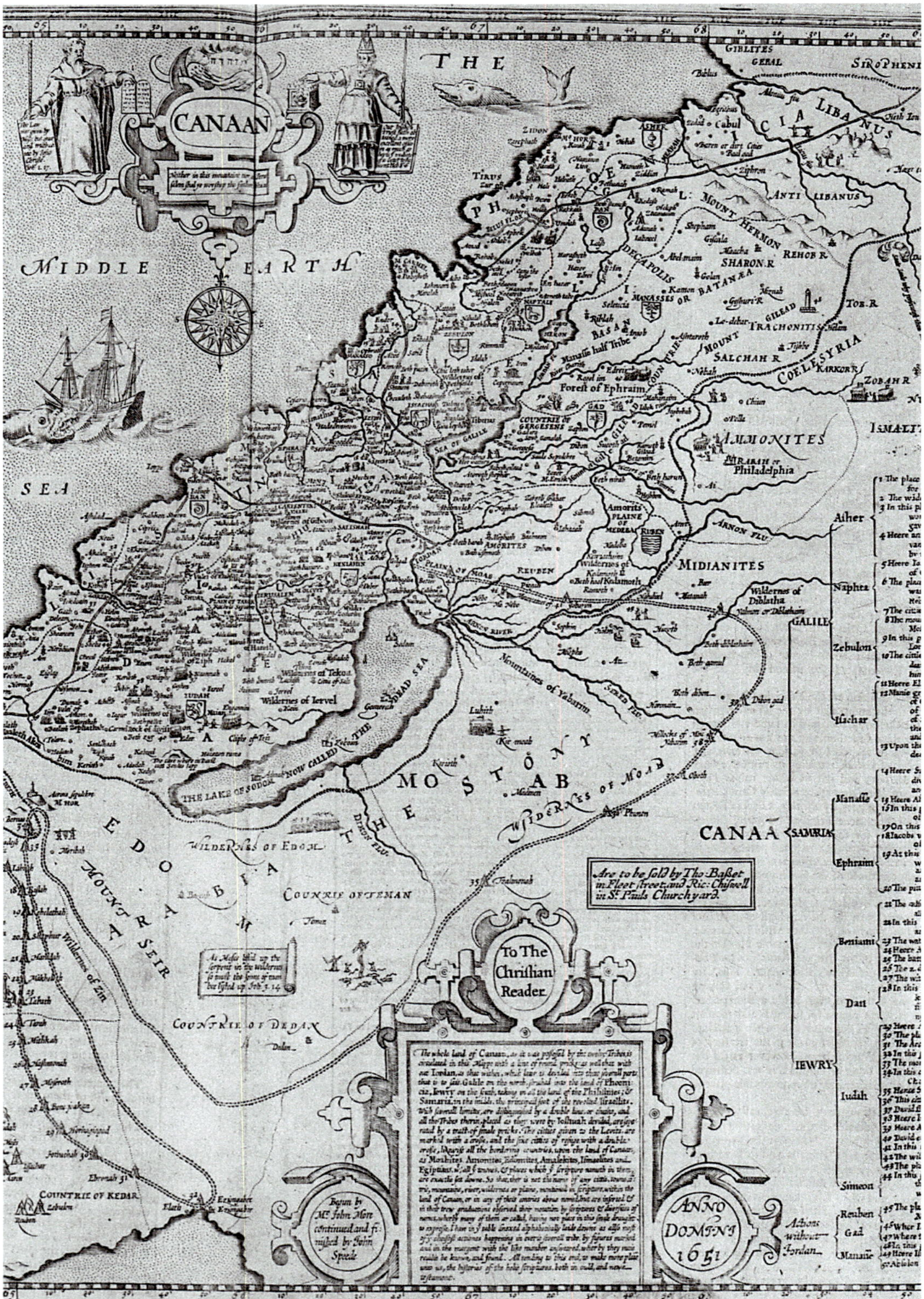

FIGURE 24. J. More and J. Speed, 'Canaan' [1611 redated 1651] (Speed 1676). (*The Board of Trinity College, Dublin*)

Dimon, Salcha, Zaphon) Speed may have gone back to Montanus. However, Speed has concentrated on biblical place-names, ignoring classical and medieval names. Speed's ultimate source is the Bible. In a cartouche headed 'To the Christian Reader', Speed makes clear that he is here presenting 'the whole land of Canaan' together with surrounding peoples such as the Phoenicians and Philistines, the Israelite tribal boundaries, and the Levitical cities and the cities of refuge;

> Likewise all the bordering countries, upon the land of Canaan, as Moabites, Amonites, Edomites, Amalekites, Ismaelites, and Egiptians, with all ye townes and places which ye scripture nameth in them, are exactlie set down. So that, ther is not the name of any cittie, towne countrie, mountaine, river, wilderness or plaine, mentioned in scripture, within the land of Canaan, or in any of their contries above named, but are inserted and in their trew graduations observed, their notation by scriptures and diversities of names, whereby many of them are called, having not place in this finale draught to expresse.

There is no indication that Speed was concerned with other sources than biblical apart from an entry which appears to read 'Gadara Strab', which may perhaps refer to Strabo, who mentions a Gadaris (Strabo, *Geog.* 16.2.45). The note on the bottom of the map which indicates that it was 'begun by Mr John More and continued and finished by John Speed' leaves one asking what More and Speed respectively contributed. We know that John More (who died in 1592) had strongly Puritan sympathies; he was a scholar and then Fellow (1568–72) of Christ's College, Cambridge, known as a supporter of the ejected Lady Margaret Professor Thomas Arkwright (Peile 1900, 78). In 1572 More moved to St Andrew's, Norwich, where he was known as a Puritan preacher. It seems likely that the strongly biblical emphasis of the map derived from More's initial work. This may explain why the learned and worthy Dr Thomas Fuller (1608–61), in his *The History of the University of Cambridge from the Conquest to 1634* (Cambridge: Deighton, 1655, ed. Prickett and Wright, 1840), says in praise of 'John More, preacher in Norwich', that 'he made the excellent map of the land of Palestine' (ibid., 184); Fuller would have approved of the biblical scholarship. It may also be (since Speed's first map favoured Arias' format) that it was More's map which drew on the Stella-Ortelius tradition, with the Palestinian coast running diagonally across the page, but used it imaginatively, 'improving' the coastline of Palestine and re-shaping the Dead Sea and re-grouping the rivers. The Puritan scholar John More, then, probably had a major influence on Speed's second map, and it is likely that his contribution lay more in the realms of biblical than cartographic expertise.

The biblical flavour is strongly emphasised by Speed's incorporation in traditional manner of the *mansiones* of Israel's journey through the desert, of vignettes (showing the crossing of the Red Sea, the cloud accompanying the Israelites in the wilderness, the battle at Rephidim, Moses on Mt Sinai, the tetragrammaton, Aaron's sepulchre, the serpent in the wilderness, the siege of Rabbath Ammon, the death of Absalom in the forest of Ephraim), and of a numbered list, set on the right hand edge of the map, of locations and associated biblical events (where Balaam cursed the Israelites, where Jacob wrestled with the angel, where the herd of swine perished, where Jacob and Laban erected a monument, where Israel defeated Og king of Bashan, and where Absalom died hanging in a tree). Speed makes his intentions clear in the cartouche: 'All tending to this end, to make more plaine unto us, the histories of the holy scriptures, both in ould, and newe testament.'

THOMAS FULLER (1608–61)

Speed's 1611 map of Canaan was republished in 1651, perhaps in response to the Revd Dr Thomas Fuller's publication in 1650 of *A Pisgah-Sight of Palestine and the Confines Thereof* in London. Fuller, a former student of Queens' College, and then of Sidney Sussex College, Cambridge, and during the Civil War chaplain to the royalist general Sir Ralph Hopton, was an Anglican preacher, writer and historian who in 1649 was given the curacy of Waltham Abbey in Essex and who in spite of his known political sympathies kept his place and his head in difficult times. He probably owed his survival to his droll humour, his personal charm, and his moderate views. W. Addison (1951, 228) describes his style as 'that of a conversationalist rather than that of a writer'. Introducing his readers in chapter 1 to the general description of Judaea, Fuller defends himself against the charges that this description of the country '(1) hath formerly been done by many. (2) Cannot perfectly be done by any. (3) If exactly done, is altogether useless, and may be somewhat superstitious'. Fuller responds that, first, 'none have formerly in any tongue (much less in English) presented us with distinct maps and inscriptions together'; secondly, while noting that 'the bowels of the best maps are puffed up with the humours of fancy, and a sceptical windiness', 'as for the differences between geographers, they ought not to make us careless to follow any, but careful to choose the best'; and thirdly, that 'though these studies are not essential to salvation, yet they are ornamental to accomplish men with knowledge, contributing much to the true understanding of the Bible' (pp. 4–5). In an age when it was important for a preacher to take the Bible with utmost seriousness, Fuller could not be faulted for producing a book which was at once so deeply rooted in study of the Bible and so careful to avoid theological controversy. With the aid of four different artists and engravers (J. Goddard, W. Marshall, T. Cross and R. Vaughan) Fuller produced 21 different maps in his volume: they comprised a map of ancient Canaan, a key map showing Galilee, Samaria and Judah, and maps showing the tribal and other regions (including one map showing Midian, Moab, Ammon and Edom, and another showing Egypt). The Transjordanian regions are covered by his main map, and the tribal maps of Reuben, Gad, Transjordanian Manasse, and the maps of Midian, Moab, Ammon and Edom, and the Desert of Paran.

Fuller's maps (Figs 25 and 26) contain a little over 200 place-names for the Transjordanian regions, apparently drawn directly from those on the map of Adrichomius published some 60 years earlier. The names of various peoples — Gergaseni, Ishmaelitae, Midianitae — and regions — Terra Ammon, Terra Argob, Terra Edom, Pars Arabia Foelicis, etc. — and references to mountains (*mons*, *montes*), rivers (*flu*[*men*], *flu*[*vius*], *torrens*), cities (*civitates*) and roads (*viae*) are in Latin, but otherwise the place-names are given in the English form found in the King James Bible, published in 1611. Nebenzahl notes (1986, p. 128) that 'Fuller supplements the delineation of Adrichom with information from his own wide reading'. Bailey (*Life*, p. 483–84) notes Fuller's 'extensive reading and researches' in Josephus, the rabbis, Pliny, Jerome, Ptolemaeus, Villepandus, Arias Montanus, Adricomius, Brocardus, Breitenbach, and especially George Sandys, who travelled in the Holy Land in 1610 and published his account of it in 1615. Fuller himself spoke of his particular debt to Brocardus: 'In setting down of their [i.e., traders with Tyre] severall places, I have wholly followed (let my candle go out in a stink, when I refuse to confess from whom I have lighted it) Bochartus in his *holy Geography*.'

Clearly, Fuller knew Adrichom's map, and used Adrichom for his place-names; but a careful comparison of the delineation of the two maps shows that Fuller has not simply copied his outlines or his location of places from Adrichom. Fuller has re-oriented Adrichom's map, and has changed the shape of the Palestinian coast-line. Fuller's Dead Sea is more rectangular than Adrichom's (though both share a small westward-turning southern foot), and the river systems differ (in particular Fuller's Jabok enters the Jordan well south of the Sea of Galilee, and there is no river entering the Jordan from the east midway along its course between the two seas, as there is on Adrichom's map). One immediately suspects that Fuller has been influenced by Speed's 1595 or 1611 map (cf. Bailey 1874, 473, 'Speed's maps seem to have been the models'), but comparison equally immediately disproves this. It is true that, like Speed, Fuller has the Dimon entering the south-east coast of the Dead Sea, but he does not join with Speed in linking the Zered to the Arnon at the northeast corner of the Dead Sea, and in general outline there is little resemblance to Speed's 1595 map, and even less to Speed's 1611 map. The precise origin of Fuller's outlines is not clear. In locating biblical sites, Fuller, if he uses Speed, does not follow him closely; to take but one example, Speed locates Rogelim and Edraei well up the River Cherith northeast of the Sea of Galilee, while Fuller sets them further south of the Cherith and directly east of the Sea of Galilee. Fuller's Index contains a list of coordinates for many of the biblical sites; it is

FIGURE 25. T. Fuller, section of map from *A Pisgah-Sight of Palestine* (1650), showing the region of Moab and the Dead Sea. (*The Board of Trinity College, Dublin*)

FIGURE 26. T. Fuller, section of map from *A Pisgah-Sight of Palestine* (1650), showing the region east of the Sea of Galilee. (*The Board of Trinity College, Dublin*)

instructive to compare his coordinates with those given by Speed on his 1595 map and by Ziegler in his book of 1536. Thus (longitude/latitude):

	ZIEGLER 1536	*SPEED 1595*	*FULLER 1650*
Edraei	66.51/32.21	72.20/35.40	69.50/33.40
Jabesh Gilead	66.55/32.21	73.00/35.30	69.50/32.50
Mahanaim	66.44/32.19	71.22/35.00	70.20/33.10
Penuel	66.39/32.18	72.20/34.20	70.10/33.00
Heshbon	66.32/32.5	72.20/33.00	70.00/32.40
Medeba	65.31/32.3	73.00/33.00	[70.20/32.20 on map]
Mephaath	66.28/31.59	72.40/34.00	69.50/32.20
Lasa/Callirhoe	66.24/31.55	71.00/31.20	70.00/31/40

The important figures to observe are the figures for latitude, for as Fuller observes of the longitudes of his predecessors, 'oh! the difference of best authors herein ... More is the certainty of latitudes'. Speed and Fuller agree against Ziegler in setting Mahanaim further north than Jabesh Gilead, but otherwise the three agree on the relative latitudes of the several places. However, on his 1595 map Speed had set the Sea of Galilee at somewhat above 35 degrees of latitude; on his 1611 map the Sea of Galilee was set about 33.20–30 degrees of latitude, and similarly on Fuller's map. Speed's 1611 map shows Canaan extending in a south-west/north-easterly direction, while Fuller's 1650 map shows Palestine in a much more northerly-southerly stance, and compresses the longitudinal axis.

Fuller was a biblical scholar operating under Cromwell's Commonwealth rather than a geographer or scientifically-minded cartographer active in Charles II's Royal Society, and his maps were prepared for the benefit of the earnest student of the Bible. Fuller knew that 'the eye will learn more in an hour from a map, than the ear can learn in a day from a discourse', and he helped his readers with an unusually large and beautifully drawn selection of illustrative vignettes (e.g., Elijah in his chariot, Moses on Mount Pisgah, the death of Absalom), and a range of symbols: thus cities of Canaanite kings were marked with crowns, conjecturally located cities with flags or banners, places mentioned only in the Apocrypha ('say not that a bar of bastardy better befitted them') with a crescent or half-moon inverted, Turkish cities with a half-moon with the points upward, and places known only from 'human writers' such as Josephus or Pliny with a cross in a circle. Asterisks indicated uncertain names. Fuller thus allows the conservative and Reformed Christian reader to focus on truly 'biblical' places, but makes the case that, though the Apocryphal writings 'were never penned by prophets, ... never written in Hebrew, never owned by the Jews, ... never solemnly quoted by Christ and His Apostles; yet because ancient ... they deserve from unprejudiced judgments a reverent respect'. The author goes on to claim that he is following the truth, 'at least the most probability grounded on Scripture, ancient authors, and modern travellers who have been eye-witnesses of the country'. Here Fuller probably had in mind especially George Sandys (1578–1644), who published his *A Relation of a Journey begun AD 1610: foure bookes, containing a description of the Turkish Empire and of Egypt, and of the Holy Land* in 1615, with a number of later editions. Sandys' work was the important precursor of a genre that became popular in the eighteenth and nineteenth centuries, and which had its effect on the development of cartography of the Holy Land.

The first 'modern' English map of the Holy Land was probably Humfray Cole's map of 1572, which was a direct copy of Stella's map published in 1570 by Ortelius. The Stella/Ortelius map was probably used by the Puritan Cambridge scholar John More, whose work was in turn used by John Speed. Speed was not primarily a biblical scholar, and he owned his debt first to Arias Montanus, and later used Adrichomius. His main source was the Bible; he seems to have had less interest in the non-biblical sources. His expressed concern was 'to make more plaine unto us the histories of the holy scriptures' — a declaration perhaps expedient amid the theological and political controversies of his time. Yet Speed was a serious cartographer, as his maps of England and Wales show, and it is hard to believe that he had no professional as distinct from simply biblical interest in the cartography of the Holy Land. Thomas Fuller, though as a clergyman under greater pressure than Speed to tread delicately, was less inhibited. He said bluntly that 'though these [cartographical] studies are not essential to salvation, yet they are ornamental to accomplish men with knowledge, contributing much to the true understanding of the Bible'. This was an important and even daring point to make, anticipating the approach of nineteenth- and twentieth-century scholarship. Fuller adapts Adrichomius' map to his purpose; he understands the importance of respecting ancient sources, he is interested in distinguishing sites evidenced from different historical periods, he notes the difficulties of calculating exactly the latitude and longitude of places, and he recognises the value of the observations of modern travellers such as Sandys. Fuller was thus perhaps the most important figure since Ziegler, but as a scholar he was to be eclipsed in due course by Adrian Reland.

PLATE I

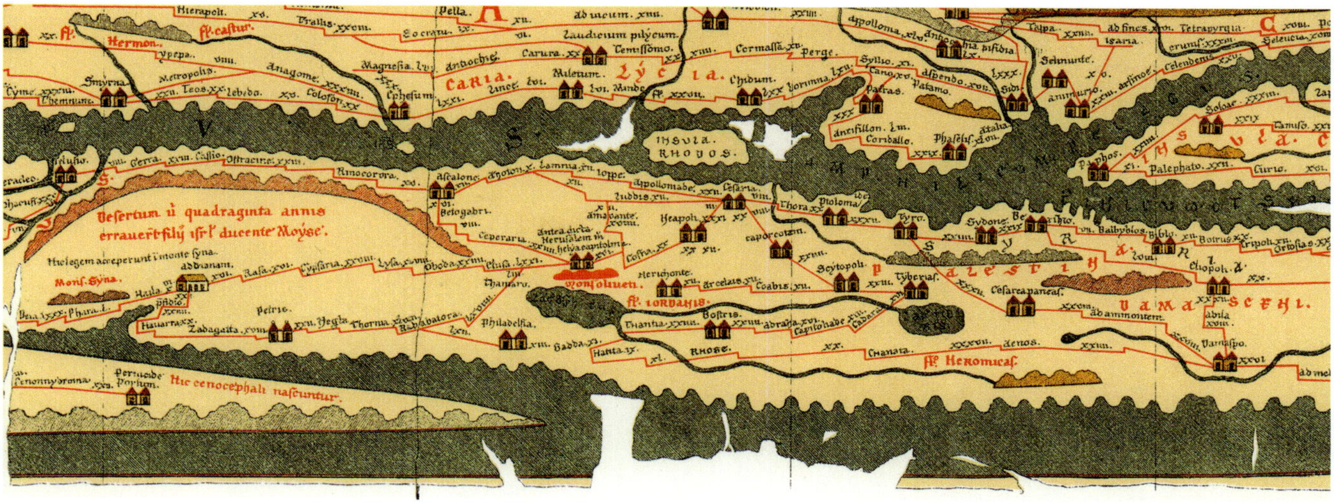

A. *Tabula Peutingeriana* (Miller 1962). (*Brockhaus/Aquarium, Stuttgart*)

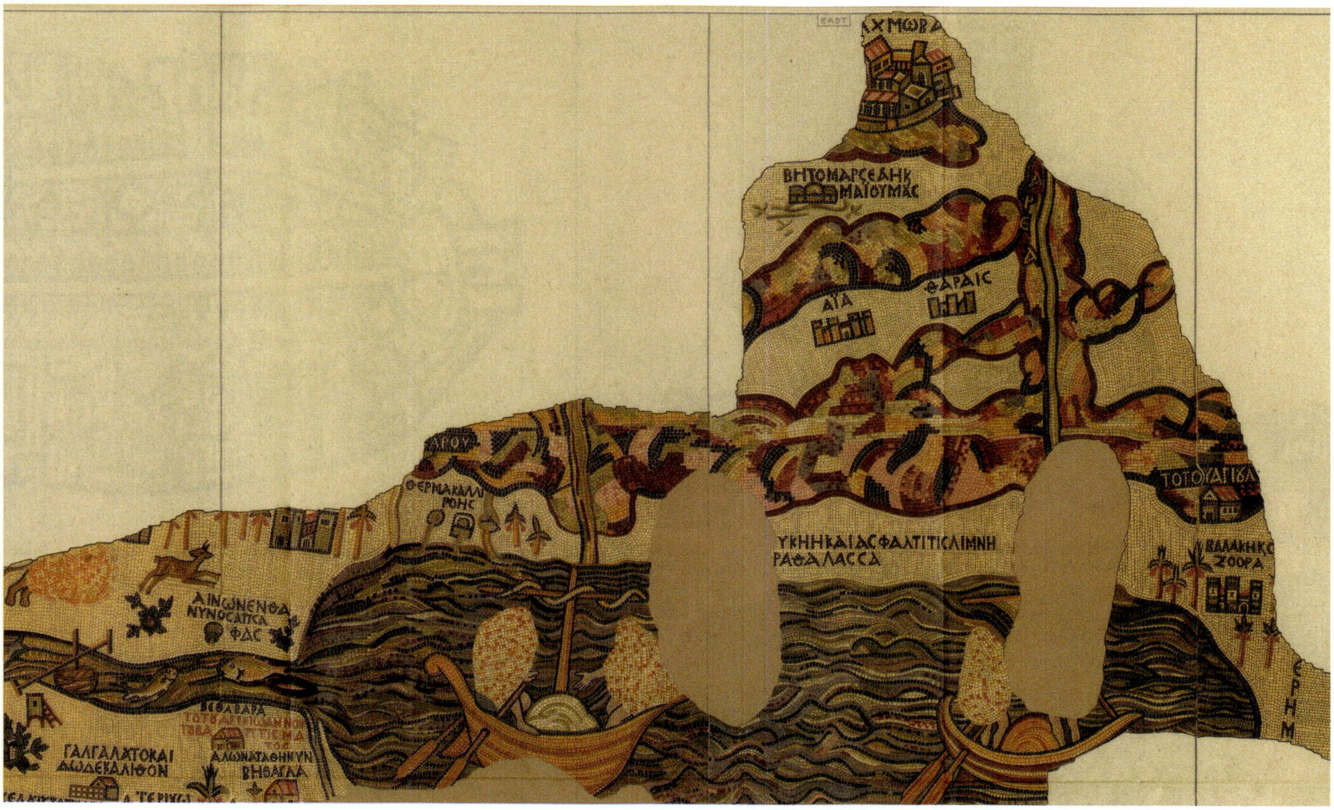

B. A section of the Madeba mosaic map, showing the region east of the Dead Sea. Note Kerak ([Ka]rachmoba) at the top (Piccirillo 1992). (*Fr M. Piccirillo and Studium Biblicum Franciscanum, Jerusalem*)

PLATE 2

Eighth-century mosaic from Umm el-Rasas, Jordan, showing (on right) seven cities, headed by Kastron Mepha'a (modern Umm el-Rasas) (Piccirillo 1992). (*Fr M. Piccirillo and Studium Biblicanum Franciscanum, Jerusalem*)

Depiction of Kastron Mepha'a in the Church of St Stephen at Umm el-Rasas (Piccirillo 1992).
(*Fr M. Piccirillo and Studium Biblicum Franciscanum, Jerusalem*)

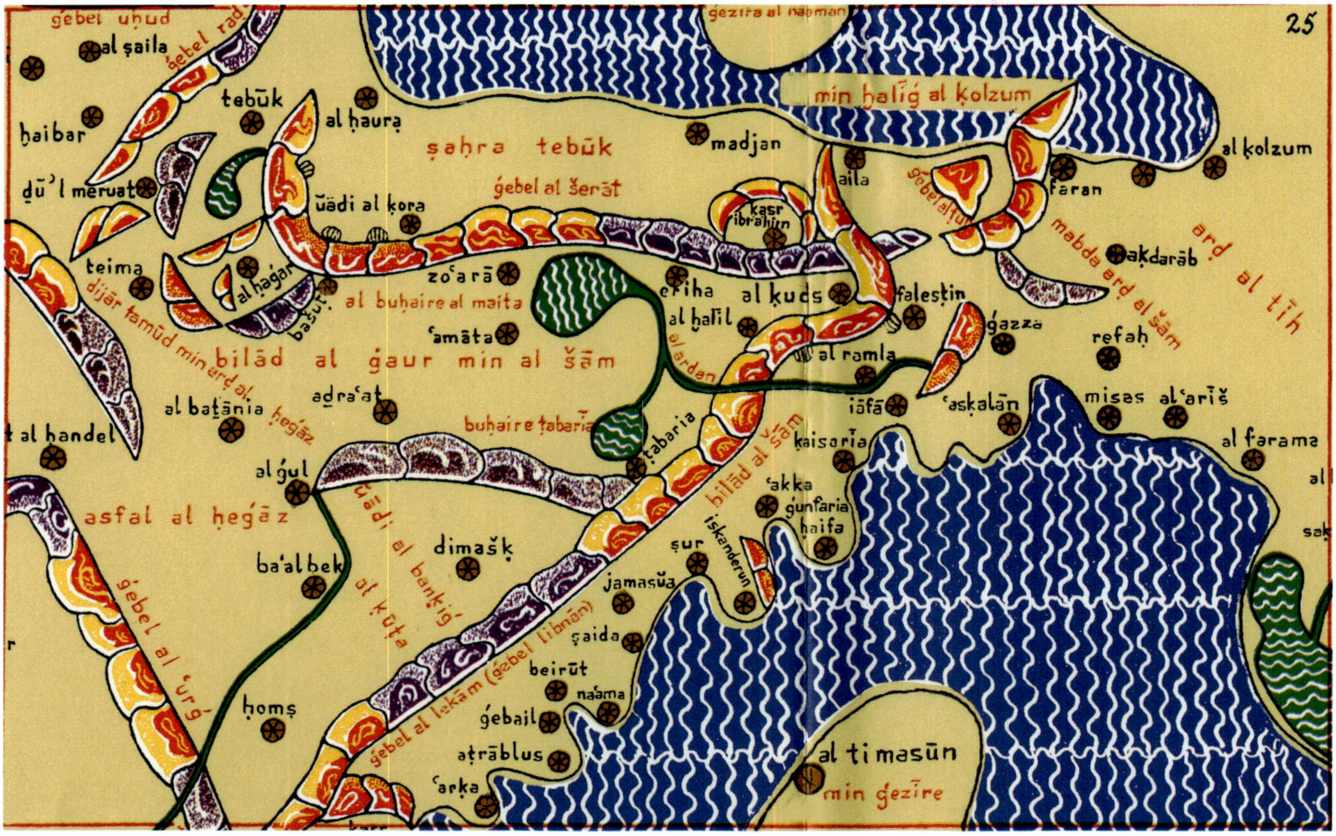

A section from Al-Idrisi's map of the world (AD 1154). The Dead Sea appears upper centre, and the Red Sea at the top (Miller 1981). (*Brockhaus/Aquarium, Stuttgart*)

Map of the Holy Land illustrating the itineraries of William Wey in 1458 and 1462: the Dead Sea (Roxburghe Club 1867).
(The Board of Trinity College, Dublin)

PLATE 6

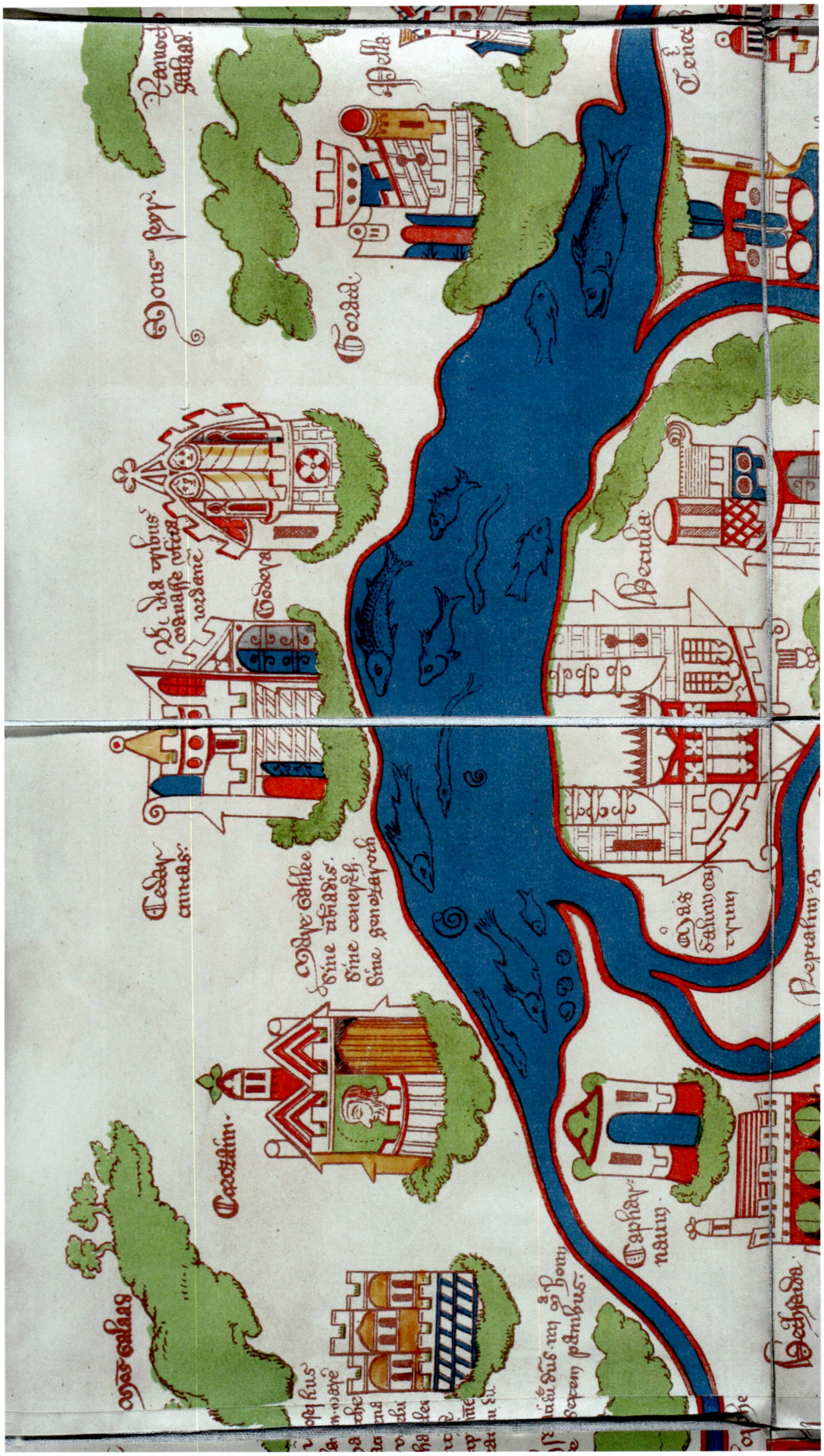

Map of the Holy Land illustrating the itineraries of William Wey: the Sea of Galilee (Roxburghe Club 1867). (*The Board of Trinity College, Dublin*)

PLATE 7

Map of the Holy Land from *Rudimentum Novitiorum* (Lübeck, 1475). (*The British Library, London*)

PLATE 8

Ptolemy, 'Tabula Moderna Terre Sancte' (northern half) (Rome, 1490). (*The Board of Trinity College, Dublin*)

Ptolemy, 'Tabula Moderna Terre Sancte' (southern half) (Rome, 1490) (*The Board of Trinity College, Dublin*)

Ptolemy, 'Quarta Asiae Pars' (Rome, 1490). (*The Board of Trinity College, Dublin*)

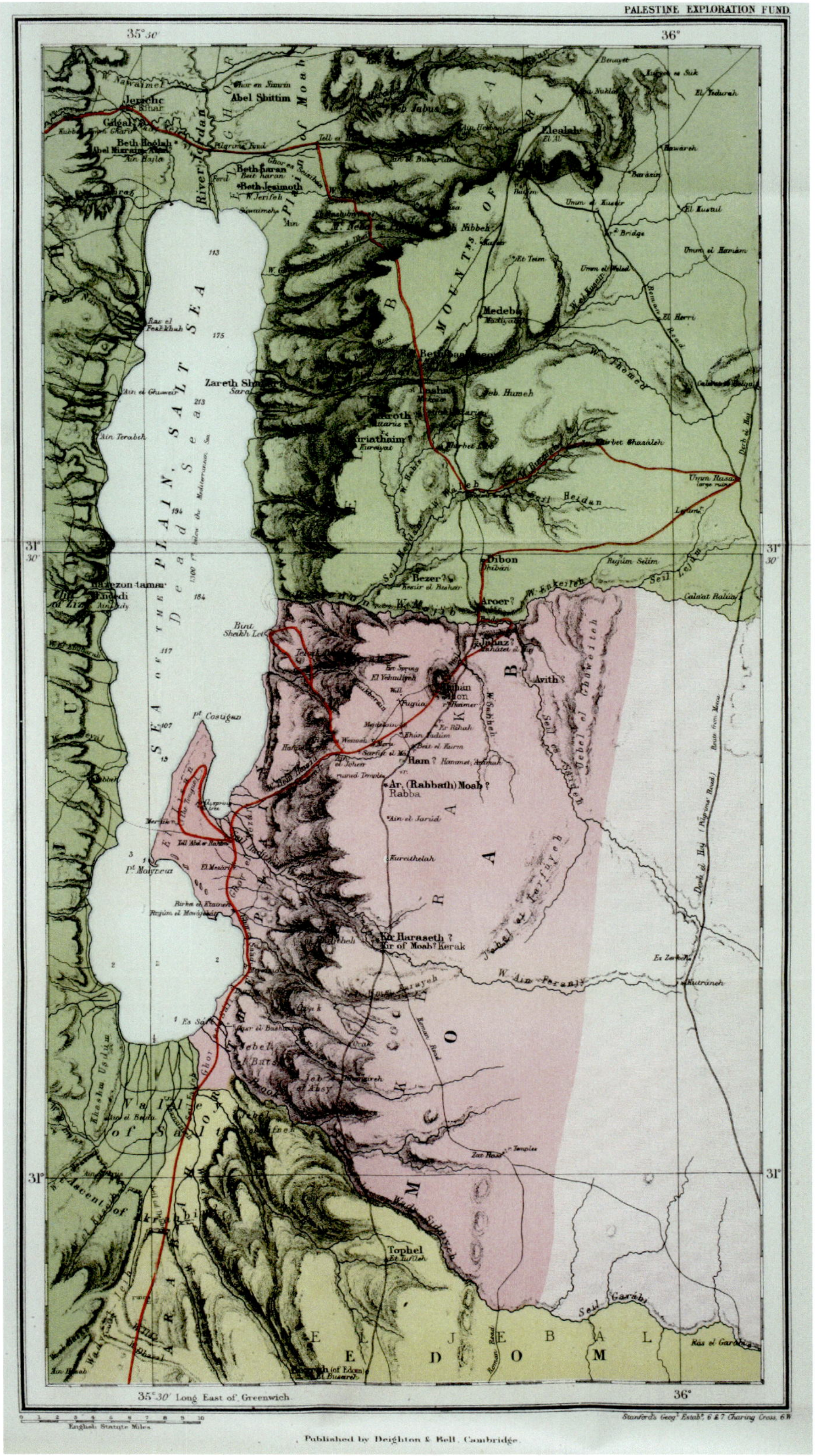

E. H. Palmer, map of Moab (*The Desert of the Exodus* I, 1871)

H. B. Tristram, 'Map of Moab' (*The Land of Moab*, 1873).

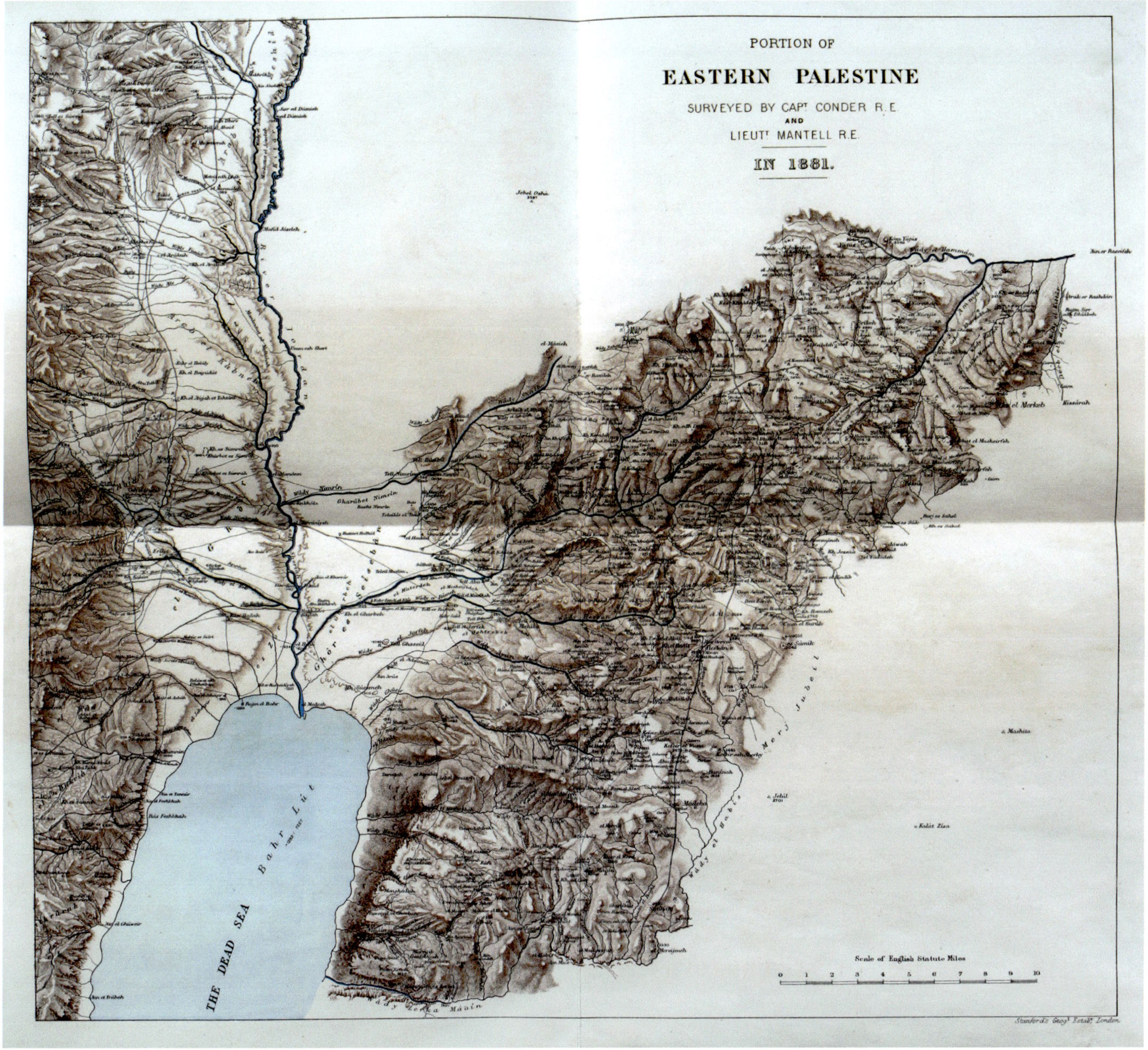

Capt. Conder, RE, and Lt Mantell, RE, 'Portion of Eastern Palestine surveyed by Capt. Conder R.E. and Lieutt. Mantell R.E. in 1881' (Conder 1889). (*The Board of Trinity College, Dublin*)

H. Fischer and H. Guthe, south-eastern section of 'Palastina 1:700,000' (ZDPV 1890, 64, fig. 3).

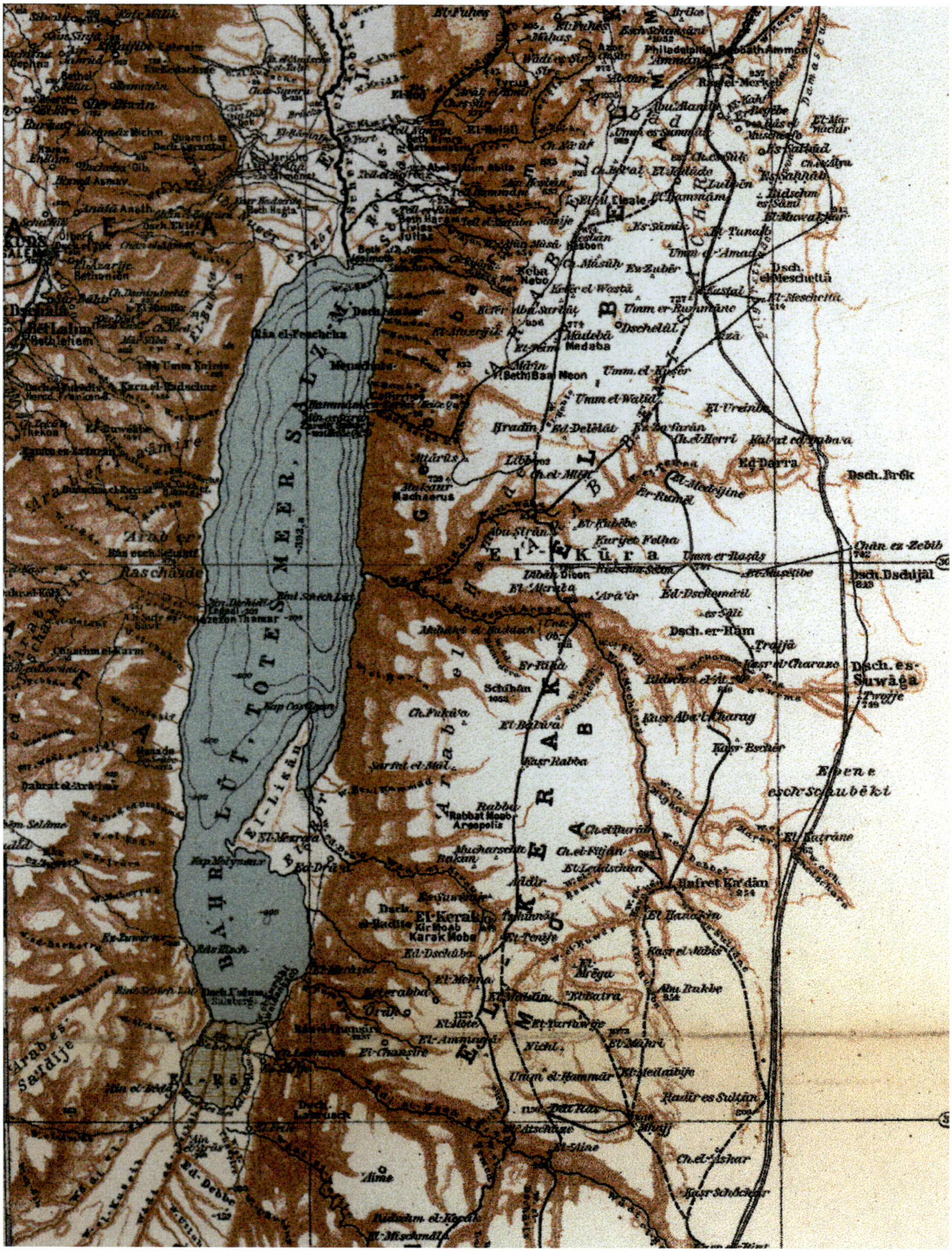

R. E. Brünnow and A. von Domaszewski, section from 'Ubersichtskarte des Ostjordanlandes 1: 700,000' (*Die Provincia Arabia* I, 1904).

Alois Musil, section from 'Karte von Arabia Petraea' (Musil, *Kuseir 'Amra*, 1907). (*The Board of Trinity College, Dublin*)

CHAPTER EIGHT

SEVENTEENTH- AND EIGHTEENTH-CENTURY TRAVELLERS IN TRANSJORDAN AND THEIR MAPS

CORNEILLE LE BRUN [BRUYN] (1652–1726)

IN 1681 CORNEILLE LE BRUN [Bruyn], born at La Haye in 1652, travelled through the Levant, returning in 1695. His *Reizen* were first published in Dutch at Delft in 1698; the French version, *Voyages au Levant et dans les principales parties de l'Asia Mineure*, was published at Delft in 1700; an English translation, *A Voyage to the Levant, or Travels in the principal Parts of Asia Minor, . . . and an account of the most considerable cities of Egypt, Syria and the Holy Land* followed in London in 1702. Le Brun illustrated his travels with a single Map of the entire Mediterranean with the Levant at its eastern end (Fig. 27). For the Jordan and Transjordanian regions, Le Brun's map shows an over-long upper River Jordan (confused with the Litani?) running into an extended Mare Galilaeae, and then following a serpentine course like that of Laicksteen's map before reaching the Dead Sea, which is slightly banana-shaped, though with an extra spur at its south-east corner. A river — presumably the Jabbok — runs into the Jordan between the two seas from the east and another — presumably the Arnon — runs into the Dead Sea at its north-eastern corner. No details are given of the Transjordan region itself. The Mare Rubrum shows no signs of any knowledge of the Gulf of 'Aqabah. Le Brun's general map did not advance near eastern cartography at all. This is not surprising, because Le Brun's travels in Palestine were limited to the regular route from Jaffa to Jerusalem, from which he visited the usual sites such as Bethlehem and Jericho, before turning north to Nazareth and Galilee. He did not cross the Jordan. The index of his book contains no reference to Ammon, Moab or Edom, or to cities like Philadelphia or Petra. In 1697 Henry Maundrell (1665–1701) made his well-known journey from Aleppo to Jerusalem, and back via Galilee, Sidon, Damascus, Baalbek to Tripoli, but he did not visit the Transjordanian region and he provided no map.

CHRISTOPHE CELLARIUS (1638–1707)

A contemporary of these travellers was the classical scholar Christophe Cellarius, professeur d'eloquence et histoire at Halle. He published his *Historia Antiqua* in Jena in 1698, and his *Notitiae Orbis Antiqui* at Leipzig in 1701. The *Notitiae orbis antiqui sive geographiae plenioris tomus alter, Asiam et Africam antiquam exponens* (Amsterdam, 1706) contains two interesting maps. The first (facing p. 282) is entitled *Palaestina seu Terra Sancta* (Fig. 28). The source of Cellarius' presentation of the Lacus Genesareth, the Jordan, and the Asphaltites Lacus is not entirely clear. The Jordan does not have the sinuous bends of Laicksteen, and the Dead Sea has not the short, tubby form of Laicksteen's sea, nor the banana-shape of Tilemann Stella's sea, nor the longer shape of Adrichom's sea with its pointed south-west corner. Cellarius' Dead Sea is more rectangular, with only the slight hint of a point in the south-west corner. The Jabok enters the Jordan about one third along its course south to the Dead Sea. The Arnon enters the Dead Sea towards the northern end, and the Zered towards the southern end, of the east shore. But most interesting is that Cellarius, following the Ptolemaic tradition, shows the Sinus Elaniticus very clearly as a definite north-eastern prong of the Red Sea, the north-west prong being identified as Lingua Sinus Arabici. For the region of

FIGURE 27. Corneille le Brun, 'Tabula geographica continens vestigia et processum itineris . . .' (*A Voyage to the Levant*, 1702). (*The Board of Trinity College, Dublin*)

FIGURE 28. Cellarius, 'Palaestina seu Terra Sancta' (*Notitiae orbis antiqui*, 1701). (*The Board of Trinity College, Dublin*)

Idumaea and Arabia Petraea, Cellarius draws most of his place names from the stations of the wilderness journey (Num. 33), but he locates the Edomite Theman and Bozra (cf. Amos 1.12) south of the Dead Sea, and Petra (identified with Arke from Josephus *Ant.* 4.82, 161) a little further south-east. A second map, between pp. 410/11, extends the view eastward under the title *Arabia Petraea et Deserta* (Fig. 29). The inclusion on this map of places like Canatha, Bostra, Philadelphia, Areopolis, Zoara, Necla, Petra, Avara, Oboda, Lysa, and Gypsaria suggests his knowledge of Latin sources like the Peutinger Table. In his preface to the reader at the beginning of Tomus II, Cellarius in fact makes plain his particular debt to Jacob Bonfrerius SJ and John Lightfoot:

Palestinam quidem, seu sanctam terram, videntur multi & verbis & delineatis tabulis descripssisse, quorum existimatione nihil ecquidem detractum volo: sed quam saepe aberraverunt a vero situ, ex ipsorum comparatione inter se, multis locis discrepantium, manifestum est, etiam ex nostro fortassis opere patebit, qui in sanctis illis regionibus, praeter antiquos, vix duce alio sumus usi quam Iacobo Bonfrerio, in Eusebii librum ab Hieronymo versum, item in Pentateuchum, Josuam et Judicum praeclaris commentariis; & Joanne Ligtfooto in Horis per Evangelistas Talmudicis.

Many seem to have described Palestine, or the Holy Land, in words and in drawn maps. I wish to take nothing from their reputation; but how often they have erred from a true position is clear when we see by comparing their work how they differ among themselves in many places. This will also perhaps be revealed from our own work, we who in those holy regions have used, apart from the ancient authors, hardly any other guides than Jacobus Bonfrerius' famous commentaries on Eusebius' book translated by Jerome, and the Pentateuch, Joshua and Judges, and John Lightfoot's *Talmudic Hours* for the Evangelists.

Bonfrerius (1575–1643) was a distinguished scholar whose 'Annotationes in Christiani Adrichomii descriptionem Terrae Sanctae' appeared in 1657 in Brian Walton's *Biblia Polyglotta*, I, 49–50. Bonfrerius' published work includes a commentary on Eusebius' *Onomasticon* (Paris, 1631; new ed. 1707). Bonfrerius' 'Tabula Geographica Terrae Sanctae', which clearly owes much to Adrichom's map, appears in N. Sanson's *Kanaan en d'omlegende Landen* (Leeuwarden: F. Halma, 1717). John Lightfoot (1602–75) was a Cambridge biblical scholar with particular interest in Hebrew and Talmudic studies; he also had assisted Walton with his *Biblia Polyglotta*. Cellarius here refers to Lightfoot's best known work *Horae Hebraicae et Talmudicae* (6 vols, 1658–78). Cellarius was a classical scholar, well versed also in the biblical texts, and at home with the works of men like Bonfrerius, Walton and Lightfoot. However, he was not primarily a cartographer; the *Biographie universelle ancienne et moderne, nouvelle edition* (ed. M. Michaud, Paris, 1844) rather disparagingly describes him as 'compilateur, et nullement géographe' (vol. 7, 309–10). Cellarius' two maps, which are not without merit, accompany his text 'De Syria' (III.12, pp. 242–83), 'De Palaestina quae et Chanaan et Terra Sancta' (III.13, pp. 283–400), and 'De Arabia' (III.14, pp. 410–32).

JEAN DE LA ROQUE (1661–1745)

Jean de la Roque was a serious traveller who published *Voyage de l'Arabie heureuse* (Amsterdam, 1716), *Voyage dans la Palestine* (Amsterdam, 1718), which unfortunately includes no map, and *Voyage de Syrie et du Mont Libanon* (Paris, 1722), which contains a sketch map of the course of the Orontes and part of Lebanon facing p. 167.

ADRIAN RELAND (1676–1718)

Far more important for our purposes was the Dutch scholar Adrian Reland from Alkmaar. He studied both classical and oriental languages with the intention of becoming a scholar, and published works on Rabbinic studies, Samaritan inscriptions, Islamic religion, and Hebrew antiquities and coins before publishing his *Palaestina ex monumentis veteribus illustrata et chartis geographicis accuratioribus ornata* (Traj. Batav., 1714). Reland was well versed in biblical scholarship, and had the additional advantage of having visited the land, in 1708. His book is a study of biblical geography, discussing geographical features, distances and site-identifications, and individual cities. Chapters relevant to Transjordan include ch. IX, De nomine Idumaeae et Phoenices; XII, De regione Edom; XVII, De Nabataeis; XVIII, De Kedarenis; XIX, De Midianitis; XX, De Moabitis; XXI, De Ammonitis; XXII, De Ituraea sive Auranitidi; XXIII,

FIGURE 29. Cellarius, 'Arabia Petraea et deserta' (*Notitiae orbis antiqua*, 1701). (*The Board of Trinity College, Dublin*)

De Trachonitidi; XXXIII, De partibus terrae Transjordanicae, Gileaditide, Peraea, Gaulanitide, Batanaea et Decapoli. There are further chapters on the seas and rivers: XXXVIII, De lacu Asphaltite; XXXIX, De lacu Tiberiadis; XL, De lacu Samochonite; XLI, De lacu Phiale; XLII, De palude Cendeuia, lacu Asphar, aliisque minoribus; XLIII, De Iordane; XLIV, De fluminibus Terrae Transjordanicae. Reland systematically examines the biblical evidence for the geography and topogaphy of Palestine; he is a worthy precursor of the twentieth-century scholars Abel, Aharoni, J. J. Simons and others. A serious modern re-appraisal of his work is overdue.

Reland accompanies his text with a set of eight maps. It is worth pausing to note what these maps present. Two of them — the 'Facies Palaestinae ex monumentis veteribus descripta ab Hadriano Relando' (facing title page) (Fig. 30) and 'Conspectus montium et camporum Palaestinae fluminumque indicia' (facing p. 305) (Fig. 31) — are in fact very similar in appearance, showing the Palestinian coast, the inland seas and the main rivers, and the mountains in relief. Very few cities or regions are named; thus, east of the Jordan divide, 'Facies Palaestinae' names only the following cities: Paneas quae et Caesarea Philippi, Bostra, Edrei quae et Adrea, Capitolias, Hippos, Gadara, Abela, Bethnabris, Chesbon, Livias, Beelmaus, Cariathaim, Medaba, Matthane, Machaerus, Petra metropolis Arabiae Petraeae et dein Palaestinae tertiae, Aila vel Elath. 'Conspectus montium . . .' names even fewer: Paneas, Chesbon, Petra. These names are predominantly from classical, not biblical sources. Reland does not try to locate the hundreds of biblical place names which fill the maps of Ziegler, Mercator, Stella, Laicksteen, Ortelius, Adrichom, and their successors. He gives the inland seas their classical names — Lacus Tiberiadis and Lacus Asphaltites. The three inland seas all assume a similar elongated shape, most unlike their shapes on the maps of Stella, Laicksteen and Adrichom and their followers. The Lacus Asphaltites shows no sign of the existence of the drowned cities of Sodom, Gomorrah, Admah and Zeboim. The Zered, a river known to the biblical rather than to the classical authors, does not appear. The Arnon enters the Lacus Asphaltites, as on many earlier maps, towards the northern end of the east shore. Most interestingly the River Jarmuk (Iarmoch fl.) appears for the first time, entering the Jordan immediately south of the Lake of Tiberias; the Iabboc enters the Jordan lower down. Reland is producing a very different map from those of his predecessors, and he emphasises its classical rather than biblical flavour by picturing Hermes, the messenger of the gods, concerned with roads, measurements, as a cartographer with compasses, maps, a globe, a surveying pole and watch, and a milepost, apparently selling a map to a winged figure, perhaps Urania, the goddess of intellectual pursuits (see Nebenzahl, 142).

A second pair of maps is concerned with the history of Palestine. The first, 'Conspectus Palaestinae uti a priscis incolis fuit inhabitata tempore Iosuae' (facing page 138) (Fig. 32), shows the peoples and regions of the land both sides of the Jordan at the time of Joshua's conquest. Terra Canaan west of the Jordan includes the Cananaei, Chittaei, Emoraei, Jebusaei, Perizzaei, and, to the north, the Chivvaei, and to the south, the Terra Edom, together with a mere handful of cities. Terra Transjordanica east of the Jordan includes the Emoraei, Moabitis (and the Campestria Moabitarum) to the south and Basan to the north, and no cities at all. What is interesting is that Reland names the seas in Hebrew: the *yam kinereth*, the *yam hagadol*, and the *yam ha'arabah*. The second map of this pair is entitled 'Conspectus Palaestinae divisae in primam, secundam et tertiam, ut erat circum initium saeculi sacrae Christanae quinti' (facing page 127) (Fig. 33). On this map, for Transjordan Reland shows only the following cities: Paneas, Capitolias, Gadara, Hippos, Esbus, Zoara, Petra metropolis, and Aila. The cartouche includes a Christian standard bearing the Chi-Rho sign, and a coin of Constantine the Great. This pair of *conspectus* shows the land before and after the Jewish occupation; Reland seems to ignore totally the political geography of the kingdoms of Judah and Israel, and the land under the major empires before the Roman empire, focusing on the pre- and post-Jewish eras.

A third pair of maps shows the distances between ancient cities as given by the ancient classical, non-biblical, authors. The first is entitled 'Facies regionis

FIGURE 30. Reland, 'Facies Palaestinae' (*Palestina ex monumentis veteribus illustrata* . . . , 1714). (*The Board of Trinity College, Dublin*)

FIGURE 31. Reland, 'Conspectus montium et camporum Palestinae fluminumque indicia' (*Palestina . . .*, 1714). (*The Board of Trinity College, Dublin*)

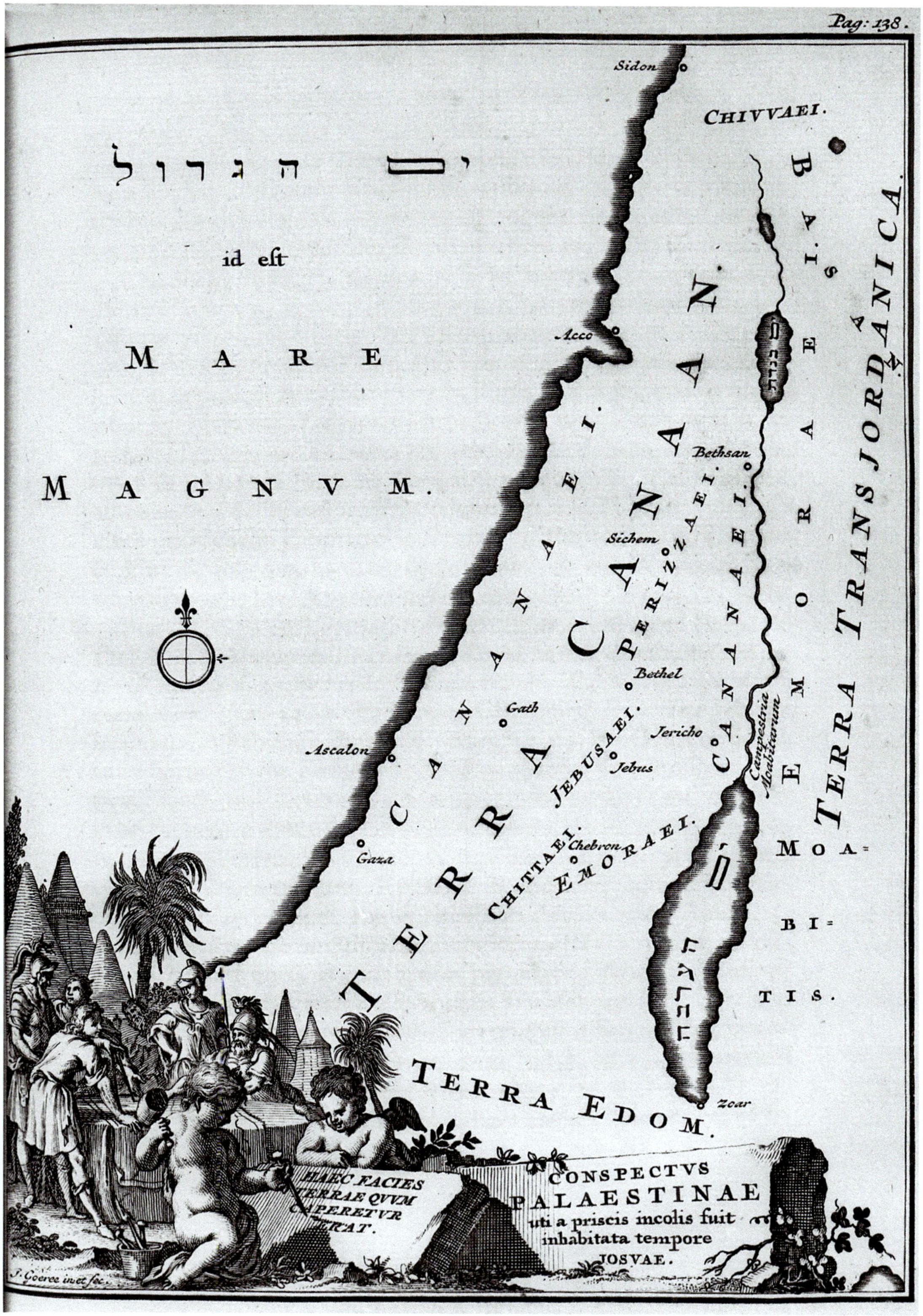

FIGURE 32. Reland, ‘Conspectus Palestinae . . . tempore Iosuae’ (*Palestina* . . . , 1714). (*The Board of Trinity College, Dublin*)

FIGURE 33. Reland, 'Conspectus Palestinae. . .ut erat circa initium seculi sacrae Christianae quinti' (*Palestina . . .*, 1714). (*The Board of Trinity College, Dublin*)

Philistaeorum in qua situs quinque urbium quas ista gens incoluit, Ecronis, Asdod, Gath, Ascalonis et Gazae testimonio scriptorum veterum confirmatus per miliaria Romana describitur' (Fig. 34). On this map, Reland locates the five Philistine cities by reference to the distances between them and other cities given by ancient authorities such as Arrian, Sozomen, Diodorus Siculus, the Antonine itinerary and the Peutinger Table; thus, for example, Gaza was 3 miles from Anthedon according to Sozomen, 3 miles from Maiumas according to Arrian, and 16 miles from Ascalon according to the Antonine Itinerary. Ascalon was 12 miles from Ashdod (Azotus) according to the Peutinger Table. (Reland's map reveals, however, that Diodorus Siculus' 34 Roman miles from the region of Gaza to Azotus do not quite fit with the other statements.) The second map of this pair, entitled 'Intervalla locorum Palaestinae digesta per miliaria Romana' (facing p. 424) (Fig. 35), again makes use of information given by the same ancient authors, together with Josephus, to locate a number of places: thus, in the Transjordanian region the Lacus Phiala is 25 miles from Paneas (Josephus), Capitolias is 26 miles from Gadara (Antonine Itinerary), Abela is 12 miles from Gadara (Eusebius), Chesbon is 20 miles from the Jordan and 6 from Mt Nebo (Eusebius), and Machaerus is 8 miles from the Dead Sea (Josephus). Petra is 37 miles from Zoar (Diodorus Siculus) and 42 miles from Elath (Eusebius). From Elath to Gaza is 157 Roman miles according to Strabo, but according to Pliny 150. Reland, in true scholarly fashion, is using maps to present information quarried from the ancient sources, which will ultimately contribute to a larger picture. This interest of Reland's in the ancient sources is underlined by his reproduction of a short section of the Peutinger Table under the title 'Conspectus Palaestinae in Tabula Veteri Itineraria'; unfortunately the section Reland has selected does not extend quite far enough to show Aila in the south or Damascus in the north. A final map (facing p.320) gives a 'Conspectus Libani atque Antilibani secundum observationes Henr. Maundrelli ex autographo eius descriptus' (Fig. 36). Maundrell's book was first published in 1698, and Reland had done well to incorporate its findings. This map shows that the Libanus and Anti-Libanus ranges run in parallel north-north-east, divided by the Litani valley; earlier maps had shown the Libanus and Antilibanus running as two parallel chains eastwards from the sea, enclosing Palestine on the north, to meet up with a row of mountains running south enclosing Palestine on the east.

THOMAS SHAW (1694–1751)

The early eighteenth century saw the development of travel in the Levant by Europeans. Le Brun, De La Roque, Henry Maundrell, and Adrian Reland were followed by the Oxford don Thomas Shaw, whose *Travels, or Geographical, Physical and Miscellaneous Observations relating to several parts of Barbary and Levant*, which included 'Observations on Syria, Phoenice and the Holy Land', was published at Oxford in 1738. Shaw's book contained 'A Map of the Coast of Syria, Phoenice, and of The Holy Land' (Fig. 37), dedicated to the Reverend John Conybeare, Dean of Christ Church Oxford, which shows the route of Shaw's journey. Apart from visiting Haifa, Joppa, Ramah, Bethlehem, Jerusalem, Bethany, and Jericho — the usual tourist trail — Shaw's visit was confined to the coastal region of Palestine, Lebanon and Syria, never reaching Transjordan, for which his map shows only Gaulonitis, Gilead, the Ammonitae, Amoritae, and Moabitae, the Fl. Jabbok, Abela, Heshbon, Mt Nebo and (Mount) Abbarim. But what was important about Reland and Shaw and their successors is that these authors could name and locate, with reasonable accuracy, the places that they had visited. This had not happened since the time of Breitenbach and Reuwich in the fifteenth century; the great cartographers of the sixteenth and seventeenth centuries did not base their maps on personal exploration but on biblical scholarship.

RICHARD POCOCKE (1704–65)

In 1737–38 Richard Pococke, an ecclesiastic who became Bishop of Ossory and then Bishop of Meath in Ireland, travelled in the east, publishing an account of his travels in two volumes a few years later under the title *A Description of the East and some Other Countries, vol. I: Observations on Egypt* (London, 1743),

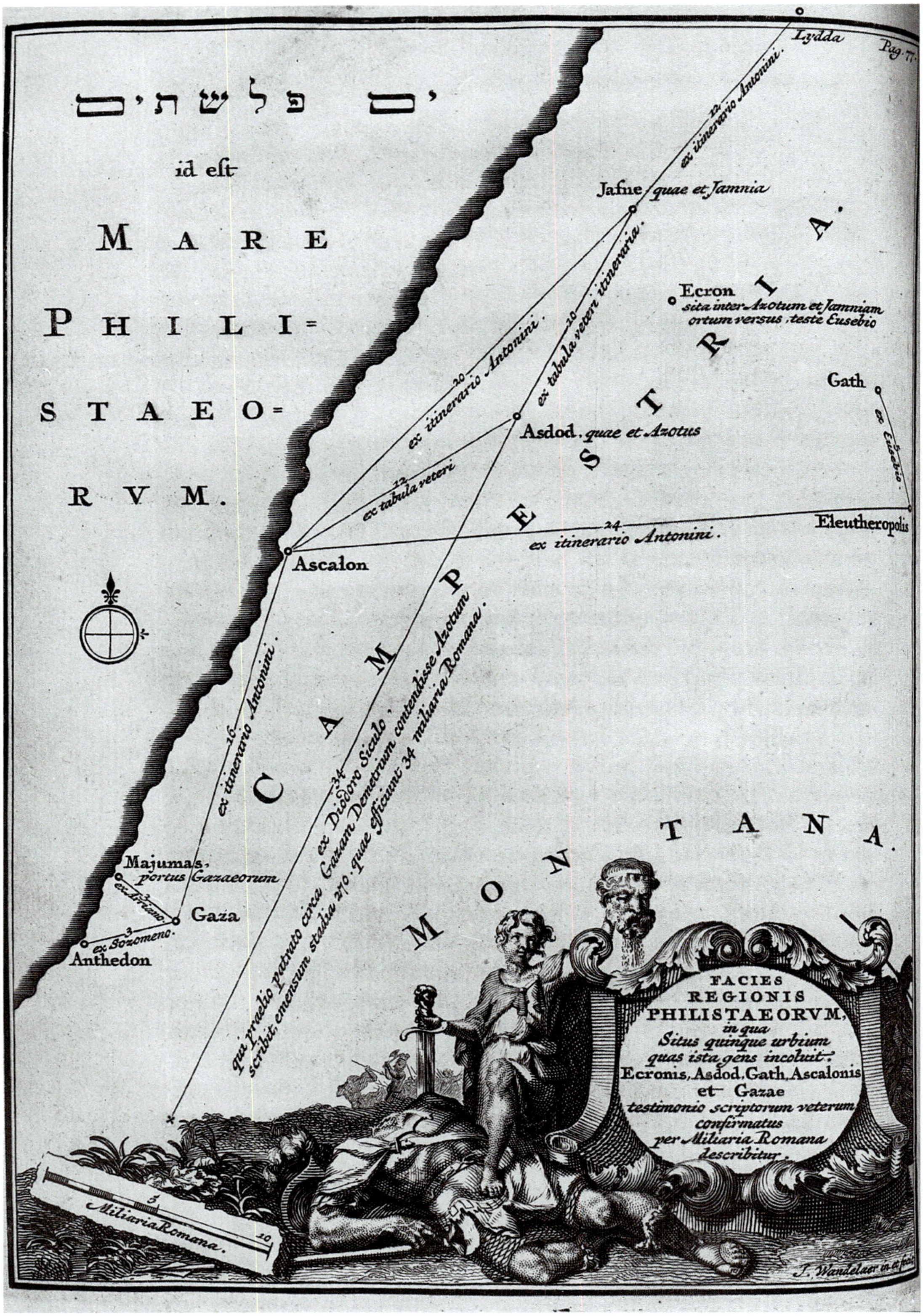

FIGURE 34. Reland, 'Facies regionis Philistaeorum . . .' (*Palestina* . . ., 1714).
(*The Board of Trinity College, Dublin*)

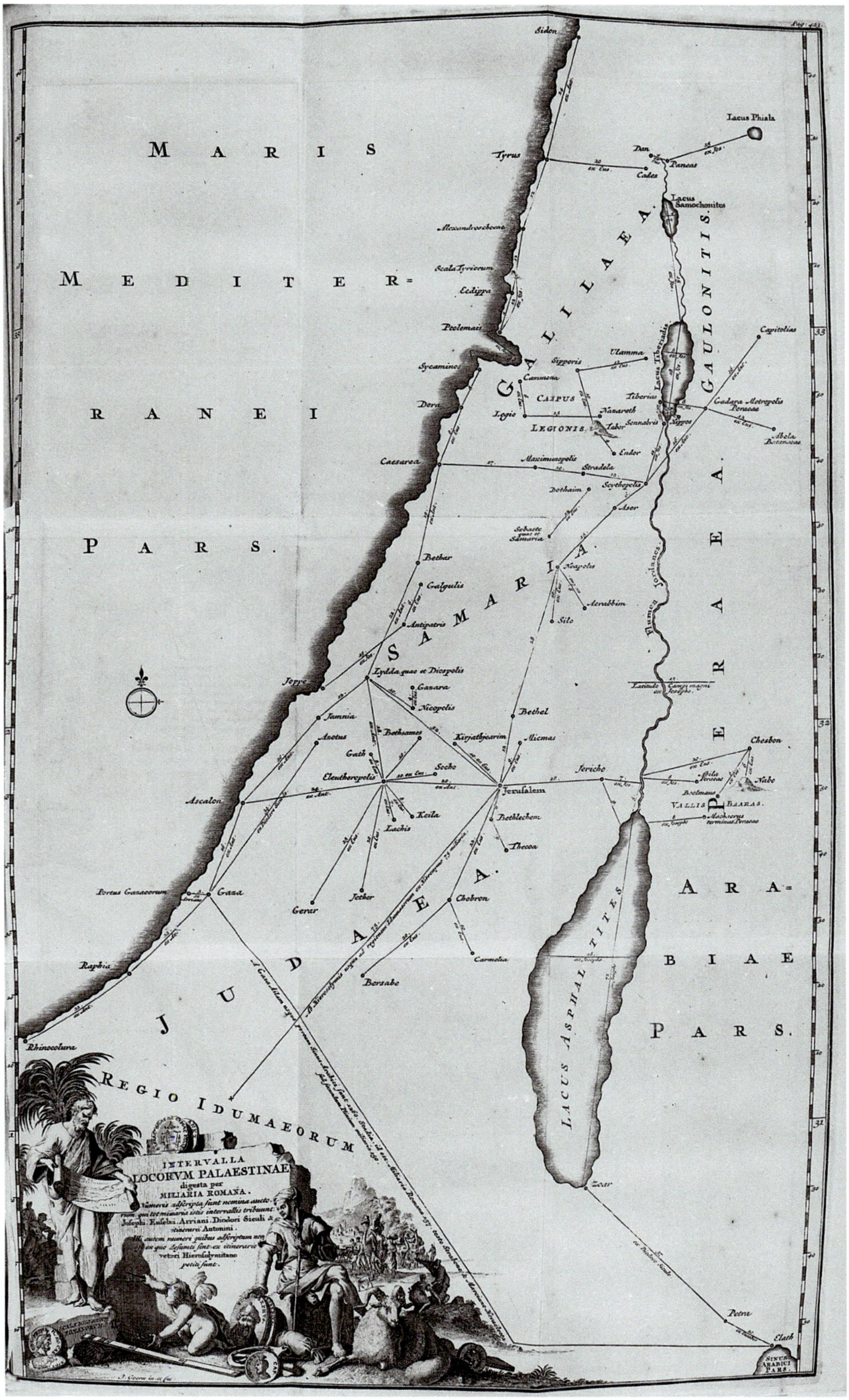

FIGURE 35. Reland, 'Intervalla locorum Palestinae digesta per miliaria Romana' (*Palestina . . .*, 1714). (*The Board of Trinity College, Dublin*)

FIGURE 36. Reland, 'Conspectus Libani atque Antilibani . . .' (*Palestina* . . ., 1714).
(*The Board of Trinity College, Dublin*)

FIGURE 37. Thomas Shaw, 'A map of the coast of Syria, Phoenice and of the Holy Land' (*Travels . . .*, 1738). (*The Board of Trinity College, Dublin*)

and vol. 2, *Observations on Palestine or the Holy Land, Syria, Mesopotamia, Cyprus and Candia* (London, 1745). Pococke similarly travelled between the coast and Jerusalem and its environs, along the coast, and in Galilee, before heading north to Lebanon. He did not cross to the Transjordan. He incorporated 'A Map of The Holy Land and Syria' (Fig. 38), drawn by the well-known cartographer T. Jefferys (*c.* 1710–71), which shows in detail the places he visited, but gives very little for the Transjordanian region. (His map of Egypt and Arabia Petraea in Vol. I reaches eastward only as far as the Egyptian coast of the Gulf of 'Aqaba.) Pococke's delineation of the inland seas suggests that he is following Reland's map. For the first time Arabic names appear on the map; thus for example L. Samachonitis is given its Arabic name Bahr el-Houly. Of rivers he shows only a small, unnamed stream entering the Jordan just south of the Sea of Galilee (presumably the modern Jarmuk), and the River Arnon, entering the northern end of the east shore of the Dead Sea. He gives a limited number of toponyms: Caesarea Philipi/Paneas, Bostra, Geser-benet-Jacob, Edrei, Julias Bethsaida/Telony, Capitolias, Gamala, Hippos, Gadara, Abela, Chesbon, Bethnabris, Livias, Beelmaus, Medaba, Cariathaim, Machaerus, M. Nebo. These names again suggest that Pococke is following Reland's map. Pococke was aware of some of the problems of biblical topography, but he is of little help to us in our exploration of the development of the mapping of Transjordan. Pococke's work was apparently known to the author or editor of *The travels of the late Charles Thompson, Esq., containing his observations on France, Italy, Turkey in Europe, the Holy Land, Arabia, Egypt, and many other parts of the world. . . , published from the Author's original manuscript, interspersed with the remarks of several other modern travellers*, which was published in Dublin in 1744. Unfortunately Thompson's 'A Modern Map of the Holy Land and the Countries Adjacent' (vol. III, p. 25) is of very little value, and of no value for the Transjordan.

JEAN-BAPTISTE BOURGUIGNON D'ANVILLE (1697–1782)

Jean-Baptiste Bourguignon D'Anville, a French cartographer and royal geographer, published his map of the Patriarchate of Jerusalem (*Patriarchatus Hierosolymitanus*) in 1732. Like Reland, D'Anville was concerned with the distances ('intervalla') between places, which he connected by lines bearing a distance in Roman miles, with an initial standing for the ancient source referred to: thus, A, Antonini Itinerarium; T, Tabula Itineraria [i.e., the Peutinger Table]; H, Hierosolymitanum Itinerarium [i.e., the Bordeaux Pilgrim], E, Eusebius et Hieronymus; J, Josephus). D'Anville is clearly basing the key points of his map of Transjordan (Fig. 39) on the Roman rather than biblical sources (appropriately enough for a map concerned with the Jerusalem patriarchate, not the Old Testament). Only a few of the better known biblical names appear, taken mainly from the biblical narratives and from Josephus rather than from the tribal lists of the book of Joshua — Iabis, Mons Galaad, Rabbath-Ammon, Iazer, Medaba, M. Nebo, Mannith, Phogor, Nahaliel, Mathane, Dannaba, Aroer, Charac-moba, Dedan, Phenon, Theman, Zared torrens, together with Beelmaus, Baaras, Livias, Cariathaim, Callirhoe, Machaerus and Herodium (why Herodium is here is not clear; is it by association with Machaerus?). D'Anville's sinuous River Jordan seems to be derived from Laicksteen's, but his elongated Sea of Galilee and Dead Sea, both somewhat similarly shaped with a deep bay on the west and a foot tapering slightly to the south-east, seem original to their creator. The Hieromax flows into the south-east corner of the Sea of Galilee; the Jabbok flows north-west from Philadelphia past Pella into the Jordan north of Scythopolis; an unnamed *torrens* flows from the 'Sea of Jazer' into the Jordan half-way between the Sea of Galilee and the Dead Sea; the Arnon flows into the north-east corner of the Dead Sea, with a sharp spit of land between it and the mouth of the Jordan; the Zered flows into the middle of the eastern shore of the Dead Sea; and finally, an unnamed river runs into the southern tip of the Dead Sea from the south.

Thirty-five years later, in 1767, D'Anville produced a revised version of this map, entitled 'La Palestine' (reproduced in Nebenzahl 1986, 150–51). The network of Roman itineraries which forms the basic

FIGURE 38. Richard Pococke, 'A map of the Holy Land and Syria' (*A Description of the East and some other Countries* II, 1745). (*The Board of Trinity College, Dublin*)

FIGURE 39. J. B. D'Anville, 'Patriarchatus Hierosolymitanus' (1732). (*The Board of Trinity College, Dublin*)

frame for the map is printed separately in a box on the right-hand side of the map, with a scale beneath linking the ancient Roman, Greek and Arabic scales to the contemporary French scale. The map covers less ground in the south than the 1732 map, stopping short at the southern end of the Dead Sea. The outline of the inland seas and rivers is the same, but D'Anville has increased the number of place-names, mainly by adding a number of biblical names (Abarim Montes, Abel Sittim, Aigaleim, Astareth Carnaim, Bethabara, Bethisimoth, Campestria Moab, Decapolis, Dibon, Eleale, Mahanaim, Salcha, Succoth and others), together with some Greek and Latin regional names (e.g., Ammonitis, Auranitis, Batanaea, Domus Zenodori, Gadarene, Golan vel Gaulon, Moabitis, Ituraea). D'Anville remains concerned to present a map with sound support from the historical sources, but as Fischer (1940, 28f.) observes, D'Anville's presentation of the Transjordan region is muddled (*verworren*); 'Gerasa liegt nordöstlich vom Tiberiassee, Canatha südlich statt nördlich vom Hieromax, Pella am Jabbok, Dibon, Madaba, Kallirhoe, Nebo und die Mundung des Arnon zu weit nördlich'.

The eighteenth century shows two important signs of development and improvement in the cartography of the Holy Land, and promise for the cartography of Transjordan. First, scholars like Cellarius, Reland, Pococke and D'Anville focused their work more on the Classical and later historical sources and less on the Bible; Reland, for example, makes no attempt to locate on his maps the hundreds of biblical toponyms to which Ziegler and Fuller had given so much attention (though Reland was a fine scholar, totally familiar with the problems of biblical topography). Secondly, the increasing availability of eastern travel to the educated aristocracy of Europe led to the development of a new genre of publication, exemplified by the writings of men like Maundrell, Shaw, and Pococke, who were interested as much in the landscape, the towns and villages, the people, and the flora and fauna of the present as in the ruins and evidences of the past. Travellers like Pococke marked on their maps places they had seen for themselves, noted the distances between them, and gave them the Arabic names they had learned from their guides. Their maps were no longer limited to biblical, or even classical, place-names; they began to show the land as modern travellers actually observed it. Unfortunately, however, few if any eighteenth-century travellers crossed the Jordan and reported back on the land beyond it; that was to be the contribution of the nineteenth century.

CHAPTER NINE

NINETEENTH-CENTURY EXPLORATION AND MAPPING OF TRANSJORDAN

ULRICH JASPER SEETZEN

MODERN EXPLORATION AND MAPPING of the Transjordan region began in the early nineteenth century. The first modern European traveller to leave a record of his journey in these parts was Ulrich Jasper Seetzen (1767–1811), who had studied medicine and natural sciences at Göttingen. But under the influence of his teacher J. F. Blumenbach and his contemporary Alexander von Humboldt, Seetzen began to interest himself in geography and exploration, at first intending exploration in Africa. In 1805 Seetzen was in Aleppo in Syria, learning Arabic. On 19 January 1806 he left Damascus and travelled south via Paneion, the upper Jordan, the west coast of the Sea of Galilee to Tiberias, and then via Edrei, Jerash, Salt (with an excursion to Amman), Eleale, Madaba, Dibân, Karak, and then round the southern end of the Dead Sea to Bethlehem and Jerusalem, which he reached on 6 April 1806. In 1810 the Palestine Association of London published an English translation of Seetzen's personal account of this journey, originally written as part of a letter, dated 16 June 1806, to M. de Zach, Grand Marechal de la Cour de Saxe Gotha. With this account was published a map (Fig. 40) showing Seetzen's route; the map's basic outlines are quite clearly copied directly from D'Anville's, though without any attempt at indicating geographical features apart from the seas and rivers. The river entering the southern end of the Sea of Galilee, however, is named as the Jarmuck (not the Hieromax), and the Jabok is named also as the Serka. Two obvious discoveries and improvements are the discovery of Jerash, which 'can be no other than the ancient Gerasa' (p. 33), and the relocation of Dibon to a position southeast of Medaba, not north-east as on D'Anville's map (though D'Anville's position of Dibon is shown; Seetzen notes that 'it is incorrectly laid down in the map of M. Paulus' [i.e., of D'Anville, whose work H. E. G. Paulus edited at Nuremberg, 1785–93]), and south of the wadi al-Wale. Otherwise this map deviates little from D'Anville's. This map, however, was surely not Seetzen's own map, but drawn for this publication of the Palestine Association. Seetzen clearly had much more information about southern Transjordan than is given on this map, as can be seen from his journal published later by Kruse (1854–59) (cf. Bartlett 1989, 16; Goren 2003, 44–47).

Seetzen's own map appears in *Monatliche Correspondenz zur Beforderung der Erd- und Himmels- Kunde* 22 (1810), facing p. 722, and in F. Kruse, *Ulrich Jasper Seetzen's Reisen durch Syrien, Palästina, Phönicien, die Transjordan-Länder, Arabia Petraea und Unter-Aegypten*, I–IV (1854–59), vol. IV (1859). This map (Fig. 41), entitled *Charte von Palaestina reducirt aus den von den Herrn D. Seetzen an Ort und Stelle entworfenen Handzeichnungen, Gotha 1810*, is the most detailed and accurate of all maps of the Transjordan region to appear before the mid-nineteenth century, and the first since the Crusades to be based on personal exploration of the land. Seetzen's map contains many more toponyms for the northern regions of the Golan and the Hauran than it does for the regions of Gilead and Moab further south, and north of the Wadi Zerka he gives many more place-names than does Burckhardt. On Seetzen's map the rivers are given their Arabic names and located with reasonable accuracy. Thus entering the Jordan from the east we find, from north to south, the Scheriet Mandûr and its upper tributary

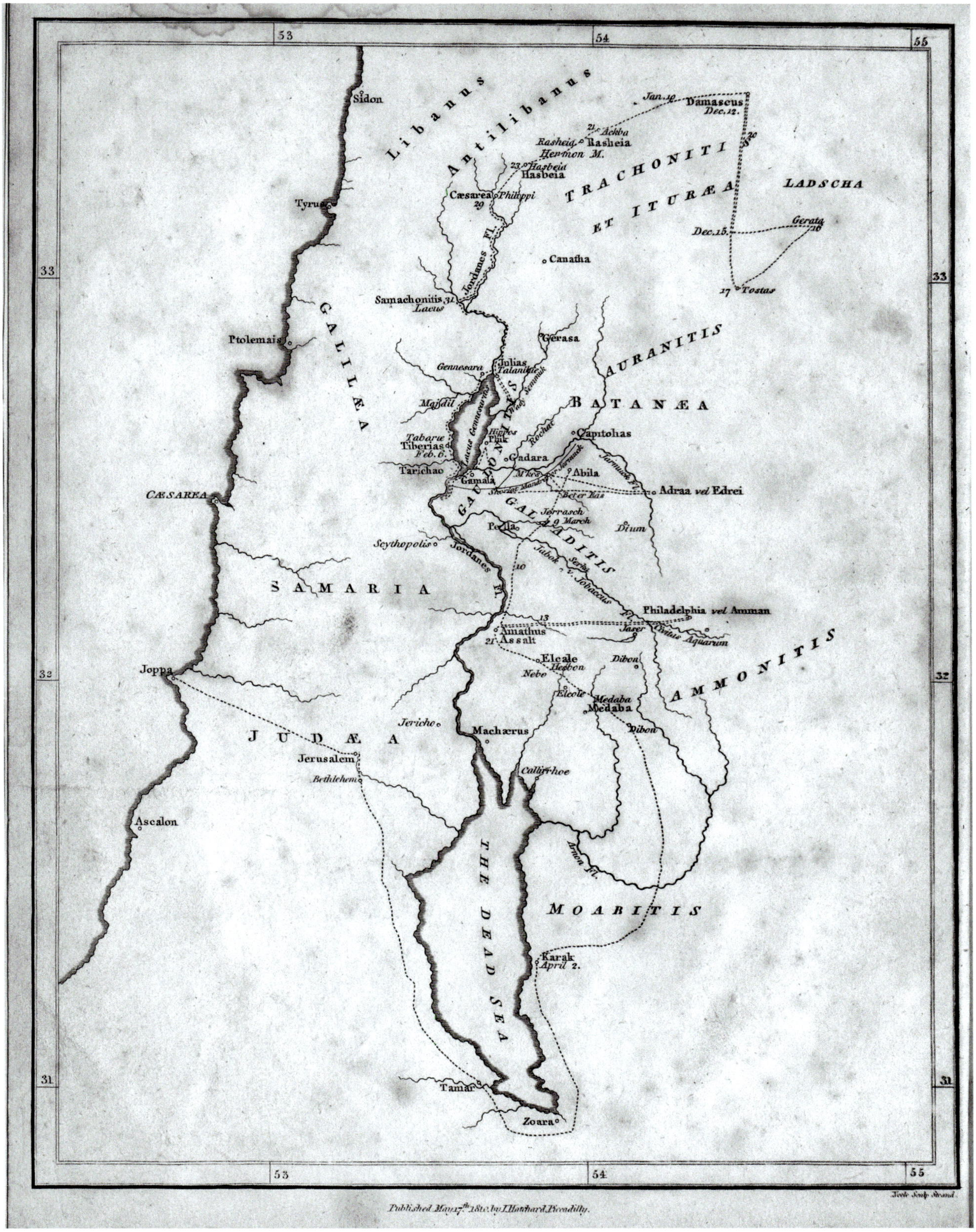

FIGURE 40. U. J. Seetzen, map from *A brief Account of the Countries adjoining the Lake of Tiberias* (1810).

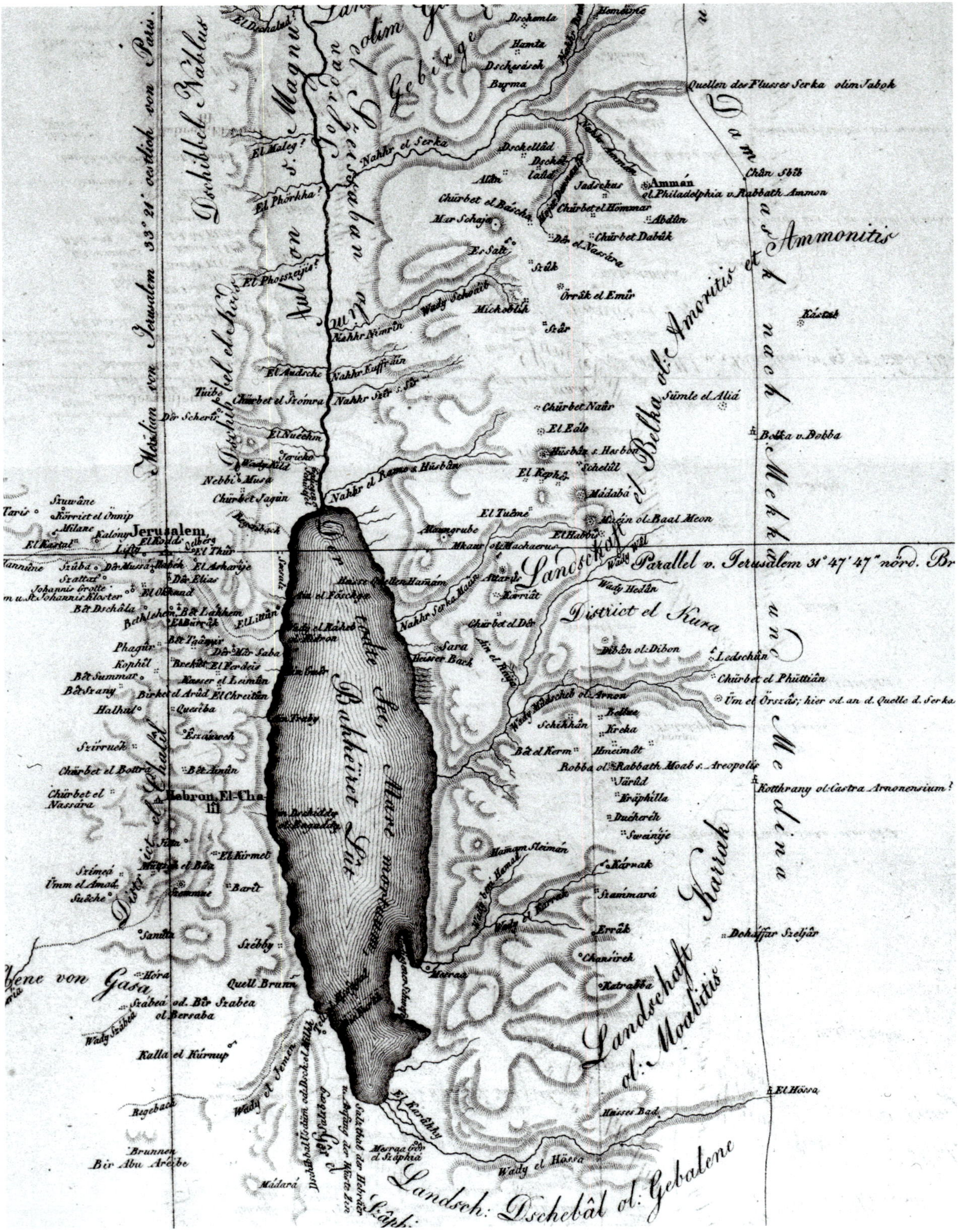

FIGURE 41. U. J. Seetzen, 'Charte von Palaestina' (*Monatliche Correspondenz*, 1810), south-eastern section. (*The British Library, London*)

the W. Rockâd (the Jarmuk), the minor W. Abu Siâd, the W. Jabes, the W. Adschlûn [= W. Kufrinjeh?], the Nahhr el Serka olim Jabok and its upper tributary the Nahhr Amman, the Nahhr Nimrîn and W. Schoaib, the Nahhr Kuffrein, the Nahhr Seir or Sîr [W. Gharaba], the Nahhr er-Rame and Husbân (entering the northern end of the Dead Sea), the Nahhr Serka Maein, the W. Wâl (W. Waleh) and W. Hedân (W. Heidan), the W.Mûdscheib olim Arnon, the W. Beni Hamad and the Wady el-Karrak, and the W. el-Hössa. Seetzen has located various places with equal accuracy, from Draa/Adraa (which he identified with Ashteroth) in the north, Dscherrasch which he identified with Gerasa), Amman (Philadelphia, Rabbath Ammon), Arâk el-Emîr, Attarûs, Medeba, Elealeh, Hesbân, Dibân, Ledschun, Mt Shihan, Balua, Rabba (Rabbath Moab, Areopolis), Karrak, Khanzireh, and Katrabba. He drew the Dead Sea more accurately than any of his predecessors, especially noting the shape of the Lisan promontory on the south-east shore of the Sea. Though he did not travel in ancient Edom south of the Wadi el-Hesa, he did through his enquiries compile lists of inhabited villages in the area, including Tafila, Shobak, Busayra and others (see Kruse 1854–59, vol. III, 17; Bartlett 1989, 16). Seetzen made an important contribution to the mapping of Transjordan. Haim Goren (2003, 46) sums up his achievement:

Er war der erste Europäer, der seit der Zeit der Kreuzritter zum Banias und Karak (el-Kerak) gelangte, er entdeckte für den Western die Ruinen von Jerash (Geras, Gerasa) und Philadelphia (Amman), spürte als erster die Quellen von Jarmuk auf und umrundete als erster Europäer das Tote Meer — eine Reise, bei der er, ebenfalls erstmals in der Neuzeit, auch Masada erwähnte.

J. L. BURCKHARDT

In 1822 John Murray's of London published J. L. Burckhardt's *Travels in Syria and the Holy Land.* The book contained an accompanying map which covered the area from the Bay of Iskanderun in the north to the Sinai peninsula and the Red Sea in the south, and from the Nile Delta in the west to Syrian desert in the east. The map (Fig. 42) shows signs of the developing knowledge of the Transjordan, especially of the Hauran region south of Damascus. Of the rivers, from north to south, the map shows two small streams entering the Sea of Galilee from the east, the Jarmuk (here the Sheriat el-Mandhour) and its upper stretch, the W. Rekad, entering the Jordan south of the Dead Sea, the W. Jabes, the Wady Zerka, the W. Sheeb, the W. Szyr, an unnamed wady (perhaps the Kefrein) between the W. Szyr and the Dead Sea, the R. Zerka Mayn, the R. Wali, which flows into the R. Medjeb (Arnon), the R. Jerra, the R. Beni Hamad,which appears to flow into a small bay which might be intended as the bay north and east of today's Lisan, a R. Dara apparently flowing down from Kerak, a R. Issal, and a Karahy Szafye or Ahsa, identified as the biblical Zered. These last three flow into the southernmost bay of the Dead Sea, south of a marked ford from east to west.

This map is noteworthy for the number of well-known Transjordanian sites it locates clearly (sometimes following Seetzen) — for example, Om Keis (Gadara), Remtha (Ramoth Gilead), Boszra (Bostra), Djerash (Gerasa), Amman (Philadelphia), Adjeloun (Ajlun), Naour (Naur), Elealeh, Heshban (Esbus), Madeba, Attarus, Diban, Aroer, Mehattet el-Haj, Beit Kerm, Rabba, Kerek, Katherabba, Khanzyre, Tafyle, Beszeyra, Gharandel (Arindela), Dhana, Wadi Mousa, Eldjy, Petra, Maan, Szadeke (Zadacata), Havana, Praesidium, Akaba, Aila, Ezion-geber. (For full details of Burckhardt's route through ancient Edom, see Burckhardt 1822, 401–04; Bartlett 1989, 16–17.) Burckhardt marks clearly the haj (pilgrimage) route running south from Damascus through Qatrany, Maan, Medawara, and Tebouk. He is the first to mark the Wadi el 'Araba on a map (van der Steen 2006, 243). The map accompanying Burckhardt's book does not do full justice to his account, but there is no doubt that Burckhardt's journey and his account of it made all the difference to subsequent mapping of the Transjordan. Burckhardt's work identified a number of places, especially in Edom, whose location was previously a matter of guesswork. Burckhardt's map, however, was far from perfect. His outline of the Dead Sea still showed some affinity with D'Anville's, and was not as accurate in some

FIGURE 42. J. L. Burckhardt, section of 'Map to accompany the travels of J. L. Burckhardt' (*Syria and the Holy Land*, 1822). (*The Board of Trinity College, Dublin*)

respects as Seetzen's (Seetzen had a much clearer picture of the Lisan), and his longitude parallels are set a little too far west, the 36 degree parallel, for example, running through Om Keis instead of some 30 km further east.

EDWARD ROBINSON, ELI SMITH, AND H. KIEPERT

In 1841 John Murray of London published the three volumes of Edward Robinson and Eli Smith's *Biblical Researches in Palestine, Mount Sinai and Arabia Petraea, a journal of travels in the year 1838*. These volumes included a set of maps by H. Kiepert. Vol. 1 contained a 'Map of the Peninsula of Mount Sinai and Arabia Petraea from the Itineraries of Robinson and Smith constructed and drawn by H. Kiepert, Berlin 1841', on a scale of 1:800,000, with an inset plan of Petra (from Laborde), and a plan of the Pass of 'Akabah. This covered the Sinai Peninsula and the Gulf of 'Aqaba, the Wadi 'Araba and Dead Sea and the land to its east to a point just north of the Wadi Mojib. Vol. 2 contained a 'Map of Palestine chiefly from the itineraries and measurements of E. Robinson and E. Smith constructed and drawn by H. Kiepert 1840', on a scale of 1:400,000. This covered the Transjordan region from the Wadi Yabes in the north to the Wadi Tufileh in the south, with the Dead Sea in full, and the corresponding land westwards to the Mediterranean. Vol. 3 contains the continuation northwards of this map, to the same scale, the Mediterranean coastline extending north of the Nahr el-Kelb, and, east of the Rift Valley, north beyond Baalbek.

These are impressive maps, showing a considerable amount of detail. For the Transjordan between Mount Hermon and 'Aqaba, Kiepert gives about 270 names (regions, geographical features, towns and villages). About 70 of these are drawn directly from Burckhardt's map. A large number of these names from modern Jordan are immediately recognisable on the 1:250,000 maps 'The Hashemite Kingdom of the Jordan' (Dept. of Lands and Surveys of the Jordan, 1949). In the Preface to Vol. 1, Robinson notes (pp. xi, xii) that

> In the parts of the country visited or seen by us, they [the maps] have been constructed almost solely from our own routes and observations and the information we were able to collect, brought into connection with known and fixed points. The other portions have been supplied from the best authorities, viz. the form and shores of the Gulfs of the Red Sea, from the chart of Capt. Moresby; the country south of Wady Musa and parts of Sinai, so far as known, from Laborde, with corrections from Burckhardt and Rüppell; . . . The small tract given of the country east of the Jordan has been reconstructed from the routes and observations of Burckhardt, compared with those of Seetzen, Irby and Mangles, and a few others of less importance . . .
>
> In the construction of the maps, it has been a main principle, to admit no name nor position on mere conjecture, nor without some sufficient positive authority. Where a place is known to exist, though its position is not definitely ascertained, it is marked as uncertain. The operation of this principle has been, to exclude a multitude of names, ancient and modern, which figure at random on most maps of Palestine.

Kiepert himself wrote a memoir on the maps accompanying these volumes, which appears as Appendix B in Vol. 3. He acknowledges his use of the maps of Berghaus, Jacotin, Seetzen, Moresby, Laborde, Ehrenberg and Bird, and of the journals of Burckhardt, Irby and Mangles, Laborde, and to a lesser extent those of Buckingham, Bertou and others, in addition, of course, to the work of Robinson and Smith.

> Of the whole region East of Wady el-'Arabah, excepting the immediate neighbourhood of Petra (which our travellers [i.e., Robinson and Smith] also visited), the southern part extending from Petra to 'Akabah is copied from the map of Laborde as the only authority; with a comparison of Burckhardt's route from Petra by way of Ŭsdakah, 'Ain Dâlegheh, and Wady Ghŭrŭndel, and thence across the western desert. The northern part, from Petra to Kerak, rests upon a new construction of Burckhardt's routes in this region, compared with the reports of Irby and Mangles. These last, though unfortunately not minute enough to admit of an independent construction, yet afford many details of importance; such as the position of the ruins Ghŭrŭndel or the ancient Arindela, and the ancient Roman road from thence to Shôbek. (p. 38)

Kiepert notes that for the regions of Kerak, el-Belka, and Jebel 'Ajlun Burckhardt's itineraries were almost the only source of information, filled out or corrected from the work of Seetzen, Irby and Mangles, and Buckingham.

> The position of Kerak, moreover, was fixed by Robinson and Smith by means of four very careful bearings, all intersecting each other in one and the same point, taken from Hebron, ez-Zuweirah el-Fôka, 'Ain Jidy, and 'Ain Terâbeh. In like manner from bearings taken by them, the mouths of all the Wadys (except Seil Jerrah) running down to the Dead Sea from the east have been determined; and also that of Wady Sha'ib in the northern Ghor. The course of the lower part of Wady Zŭrka Ma'in, with the warm springs of ancient Callirrhoë, is known by the route of Irby and Mangles thither from Ma'in, and from the bearings taken by them to various points on the west of the Dead Sea. On the same authority, the position of Ma'in, which Burckhardt gives as S.E. of Hesbân, is corrected to S.W. For the region around es-Salt, in addition to Burckhardt, the notices of Seetzen and Buckingham have been used. (p. 47)

Kiepert discusses the difficulties of calculating correct distances between places from the notes of Burckhardt (camels travel at speeds per hour varying according to the terrain), and of plotting the precise latitude and longitude of specific places from the construction of travellers' routes and bearings taken (see, e.g., his discussion of the latitude and longitude of Petra, p. 37). Kiepert gives Petra's longitude as 35 degrees, 38′ 9″ east of Greenwich, and its latitude as 30 degrees, 25′ north (p. 37); the modern coordinates are 35 degrees 29″ E, 30 degrees 19″ N. Thus

> In Jebel 'Ajlun the position of Kul'at er-Rubud in particular is corrected by the bearing of Robinson and Smith from Taiyibeh; by that of Irby and Mangles from the point where they crossed the Jordan between es-Salt and Nabulus; and also by the route of the same travellers from Beisan to Jerash. It thus comes to lie considerably further west, and somewhat further south, than on Berghaus's map. This causes again a change in the position of Jerash as laid down on the same map; and I have given it in lat. 32 degrees, 19′ 50″, approaching nearer to Moore's latitude by observation, 32 degrees 16′ 30″, while Berghaus has it still 2 minutes further north . . . (Appendix B, p. 48) (The modern coordinates are 35 degrees 44″ E, 32 degrees 20″ N.)

Kiepert's map is not based on a full trigonometric survey, but, working in the 1830s, he was well aware of the surveying methods employed at the time. His map therefore is only as accurate as the use of recorded bearings and calculated distances can make it. The marvel is that Kiepert can present a map so recognisable to modern scholars and travellers.

There are four obvious features to be observed in Kiepert's maps. The first is that, for the Transjordan region, there are only about 40 place-names derived directly from the biblical or other ancient sources. These include the well-known classical regions like Ituraea, Batanaea, Gaulanitis, Arabia Petraea, and biblical regions like Basan, Gilead, Ammon, Moab, Gebal; ancient sites like Baneas/Caesarea Philippi, Gerasa, Philadelphia, Callirrhoe, Areopolis, Thoana, Arindela, Petra, Zodocatha, Aila, and biblical sites such as Jabesh Gilead, Rabbath Ammon, Beth Nimrah, Elealeh, Heshbon, Medaba, Baal Meon, Dibon, Aroer, Kir Moab, Zoar, Tophel, Bozrah, and Hor, most of which could be easily identified with sites with similar Arabic names. Most of these names were included in maps covering this region from Reland's time onwards. The second feature is that the remaining place-names, over 300 of them, were the modern Arabic place-names, gathered and noted by travellers like Seetzen, Burckhardt, Irby and Mangles, and Buckingham. Their inclusion represents a new attitude towards cartography of the Transjordan. Robinson and Smith are specifically concerned 'to open the treasures of Biblical Geography and History still remaining in the Holy Land' (Preface, 1, xii–xiii), but they were well aware of the preservation of ancient names among the native population of the land, and therefore of the need to discover and record them. The third feature is that these maps attempt to draw the courses of the rivers and wadis of Transjordan, and to indicate the chief mountains, and to give them their Arabic names. Thus the upper tributaries of the Jordan are named, as are most of the wadis entering the Sea of Galilee and lower Jordan valley from the east: Wady es-Semak, Sherîat el-Mandhûr or Yarmuk, W. el-'Arab, W. el-Kŭseir, W. et-Taiyibeh, W. Musa, W. Yabes, W. Rajib or Ajlûn, W. Zŭrka (Jabok), W. Shaib, W. Keferein, W. Seir, Ain es-Suweimeh, W. Zŭrka Maîn, W. el-Môjib and its tributaries, including W. Wâleh and W. Salîheh, Seil Jerrah, W. Beni Hemâd, W. ed-Dera'ah or Kerak and its upper tributaries, W. Kuneiyeh, W. el-Kurāhy (es-Safieh) and W. el-Ahsy, W. Tufîleh, W. Ghŭrŭndel, W. Bŭtāhy, W. el-Jeib, W. Ghuweir, W. Sikakîn, W. Nemela, W. el-Abyad, W. Abu Kusheibeh, W. Dalegheh, W. and Ain Ghŭrŭndel, W. el-Muneileh, W. el-Ithm. Of the mountain peaks, Kiepert marks Jebel Ōsha' (cf. modern Nabi Yusha), on Mt Gilead, Jebel Attârûs just south of the Wadi Zŭrka Main, Jebel Shihân, just south of the Wadi Môjib, and Jebel Neby Harûn near Petra, with the esh-Sherah mountain range running south of it. The fourth feature of these maps is that Kiepert includes modern tracks and roads, as well as references to Roman roads. Kiepert is interested in modern as well as ancient travellers. His set of maps is altogether an impressive achievement, and a great advance upon Burckhardt's map from the same publisher 19 years earlier.

In 1852 Robinson and Smith made a further journey to Palestine and Syria, their itinerary taking them from Beirut via 'Akka to Jerusalem, and then, after several local expeditions, to Damascus and Lebanon. They did not therefore on this journey add to their knowledge of the Transjordan region. The results of this journey were published in 1856 in *Later Biblical Researches in Palestine and the adjacent Regions: a journal of travels in the Year 1852* (London: J. Murray). This volume contained 'A New Map of Palestine including also Phoenicia and Coelesyria constructed principally from unpublished materials furnished by Revd Ed. Robinson, Revd Eli Smith, and Revd J. L. Porter', on a scale of 1:600,000, in two separate sheets showing the northern and southern halves respectively (Fig. 43). However, for the third edition of Robinson and Smith's work, *Biblical Researches in Palestine and the adjacent regions: a journal of travels in the years 1838 and 1852*, published in three volumes in 1867, Kiepert drew a new map. Vol. 1 included a 'Map of Lower Egypt and the Peninsula of Sinai', on a scale of 1:1,000,000, 'chiefly from the materials furnished by E. Robinson and Eli Smith, with additions from the surveys and travels of Linant, Callier, Laborde, Wilkinson, Lepsius, Koller, Russegger, Abeken, Wallin and others'. This map covered the Sinai Peninsula and the territory east of the Gulf of 'Aqaba and the Wadi 'Araba and the Dead Sea as far north as the W. Zŭrka Maîn. The second and third volumes contained the southern and northern halves respectively of 'A New Map of Palestine', originally published in the *Later Biblical Researches* in 1856.

These maps were on a smaller scale, but sharper and clearer in presentation, and much revised from the 1841 edition. Kiepert gives about 260 names for the Transjordan region between Caesarea Philippi and 'Aqaba, between the Rift Valley and the Haj road, and of these about 150 come from his first map. Kiepert has therefore changed about two fifths of the names given on the first set of maps. A comparison of the lists shows that the changes are mainly in the names of Arabic towns or villages, especially in the northern half of the Transjordan region, north of the W. Zŭrka. (On this, cf. Goren 2003, 96: 'Auch in dieser Karte arbeitete Kiepert Korrecturen von Daten und Einzelheiten ein, in besondere in Hinblick auf die nördlichen Regionen Palästinas und des Jordantals.') The 1856 map of the southern region, south from the W. Zŭrka, shows less change from the 1841 edition, though it marks the routes of Burckhardt (1812), Baron Koller (1840) and Wallin (1845). A brief description and critique of the work of Robinson and Smith, and of Kiepert's accompanying maps, is given by Fischer [1940, 48–55]. He notes that Kiepert removed the small sea (*der kleine Jordansee*) found on Burckhardt's and Berghaus's maps between Lake Huleh and Lake Tiberias (p. 50); that Robinson and Kiepert say nothing of the island (*rudschm el-bahr*) in the Dead Sea (p. 51); that they make the terrace demarcating the W. Arabah and the Ghor much too firm (p. 52); that Kiepert's representation of the terrain betrays an untrained hand, that his distinction between seasonal and perennial watercourses is unclear and contains errors. The 1856 maps remove some errors, but still lack the upper sources of the Jordan and the spring Ain el-'Arus and the marshes at the southern end of the Dead Sea, and 'Die Terrainzeichnung ist nicht mehr als eine flüchtige, ungenaue und ausdrucklose Skizze, die den meisterhaften Schilderungen Robinsons in keiner Weise gerecht wird' (p. 54).

An important critique of the Robinson-Kiepert maps is made by J. M. Miller (2000, 351–66). Miller

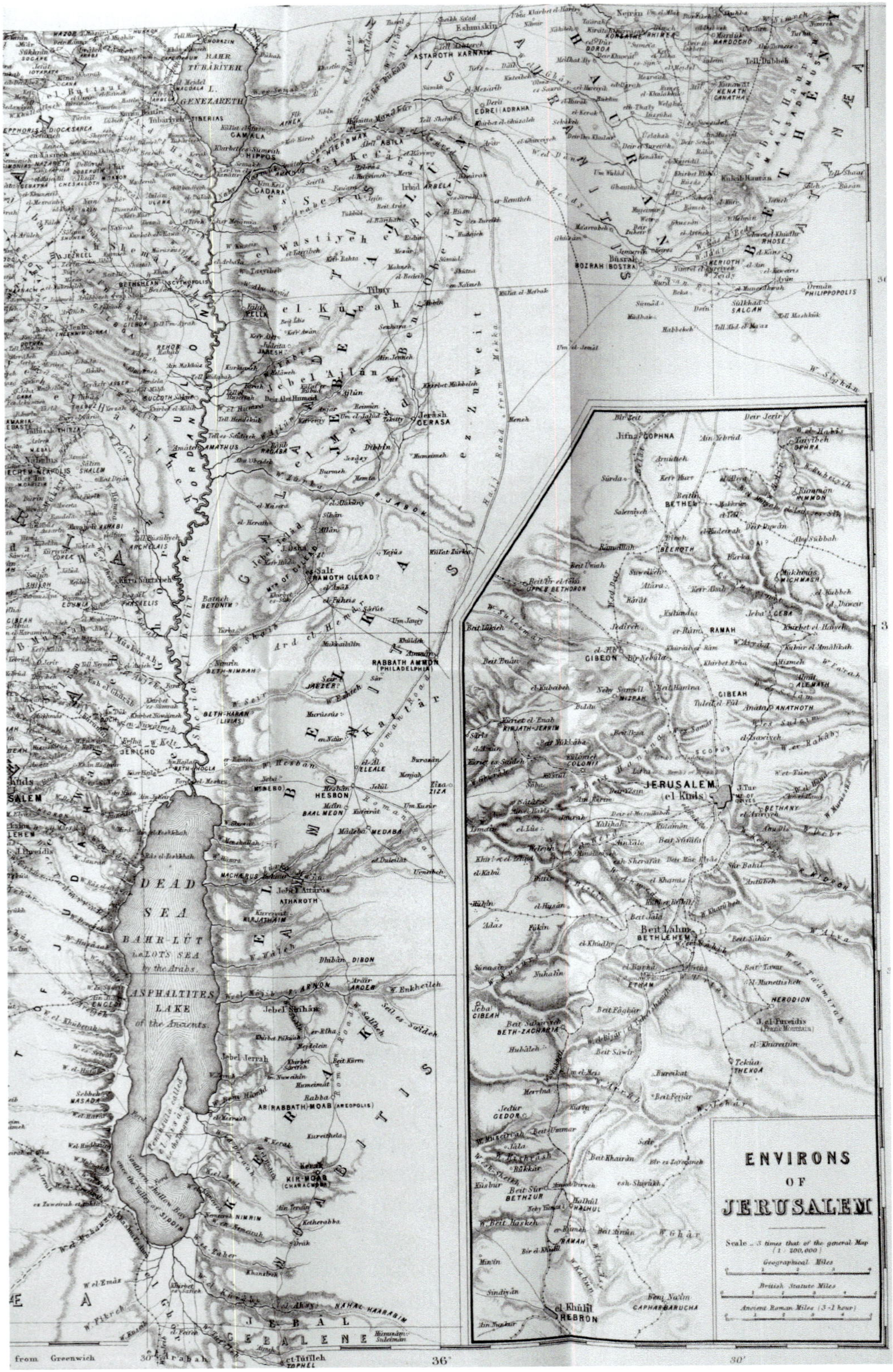

FIGURE 43. H. Kiepert, 'A new map of Palestine' (Robinson and Smith, *Later Biblical Researches*, 1856). (*The Board of Trinity College, Dublin*)

notes five features in the first version of Kiepert's map of the region south of the W. Mojib 'which can be traced either to mistaken information in Burckhardt's journal or to Robinson's misinterpretation of statements in the journal' (p. 358): (1) W. Enkheileh or Lejum approaches the Mojib from northeast, rather than from the southeast; (2) the Mojib tributary W. Saliheh is presented as a major rather than a minor branch of the Mojib, approaching it from the southeast rather than from the south; (3) Kiepert places Ariha (Araayr, El Ryhha) too far south and Jebel Shihan too far north; (4) Kiepert depicts a large, unidentifiable, bow-shaped mountain chain to the east of the sources of the Mojib; and (5) Dhat Rās appears southeast of Kerak in a setting corresponding roughly to that of Kh. el-Batra'. Miller demonstrates how these features, and resulting confusions, persisted in and affected subsequent maps, including Kiepert's second version, C. W. M. van de Velde's 'Map of the Holy Land' (1858), the map in E. H. Palmer's *The Desert of the Exodus* (vol. 2, p. 471) (1872), the map in H. B. Tristram's *The Land of Moab* (1873), and the map 'Palâstina' published by H. Fischer and H. Guthe (1890, 64, fig. 3). Kiepert's second map corrects some errors: the mountain chain east of the Mojib and several difficult sites have been removed, another major branch, the W. Shekik (W. esh-Shuqeiq) is added to the Mojib west of W es-Salihi, and from de Saulcy's work three new sites are added and the wadi system between the Lisan and the W el-Hasa is revised. Miller's work applies to just one small section of Kiepert's map; doubtless close examination of other areas would entail similar corrections.

C. W. M. VAN DE VELDE

Two years later the Dutch naval officer C. W. M. van de Velde published his 'Map of the Holy Land . . . from his own surveys in 1851 and 1852, from those made in 1841 by Majors Robe and Rochfort Scott, Lieut. Symonds and other officers of her Majesty's Corps of Royal Engineers; and from the results of researches made by Lynch, Robinson, Wilson, Burckhardt, Seetzen &c.' (Gotha: Justus Perthes, 1858). This map was in the scale of 1:315,000, and so on a larger scale that Kiepert's. The map covers the Mediterranean coast from just north of Tripolis to just south of Gaza; inland, it includes the Beka'a valley, the Hauran and the Golan, and, southwards of these regions, the Transjordanian region to the southern end of the Dead Sea. In this Transjordanian area, van de Velde's map does not reach much further east than modern Amman.

For the land west of the Jordan, van de Velde had the advantage of being able to base his map on the trigonometrical survey made by Lt J. F. A. Symonds, RE, in 1840–41 (Velde 1858, 6; Jones 1973; for a diagram of Symonds' triangulation, see ibid., 35). (Van de Velde confesses in his *Memoir* that it lay beyond his power to make a complete triangular survey, and that he carried not a theodolite but a telescope and 'a surveying instrument' which he describes as a 'the "boussole d'arpentage" with two levels, cross-threaded plunging telescope and vertical semicercle [*sic*] with nonius'. He later disconnected the vertical semicercle, finding it useless for his purpose (*Memoir*, 3–4).) However, Symonds' triangulation did not reach east of the Jordan, and for this region van de Velde had to rely on the accounts and maps of previous travellers and upon his own observations. In fact, Van de Velde made only a single day's visit (15 May 1852) east of the Jordan, crossing the Jordan eastwards just south of the wadi Yâbîs, and travelling via Kefr Abîl to Tubakât Fahel (Pella) before recrossing the Jordan westwards. Later he crossed Lebanon from the west via Jezzin, and north of Rasheiya, to Damascus. Van de Velde's own observations were limited to a small number of bearings taken on Transjordanian sites or features, e.g., the northern summit of Mt Hermon, Kŭl'at er Rŭbŭd, Fîk and several other places from various points (see *Memoir*, pp. 111, 131, 137, 142, 143, 146). Thus south of Damascus, with the exception of the immediate vicinity of Pella, van de Velde was almost totally reliant for his map (Fig. 44) on earlier sources.

The Transjordanian region, defined by the Jordan valley on the west, the Haj road on the east, and the latitudes of Banias (Caesarea Philippi) in the north and the Wadi el-Hasa in the south, contains on van de Velde's map about 320 place-names.

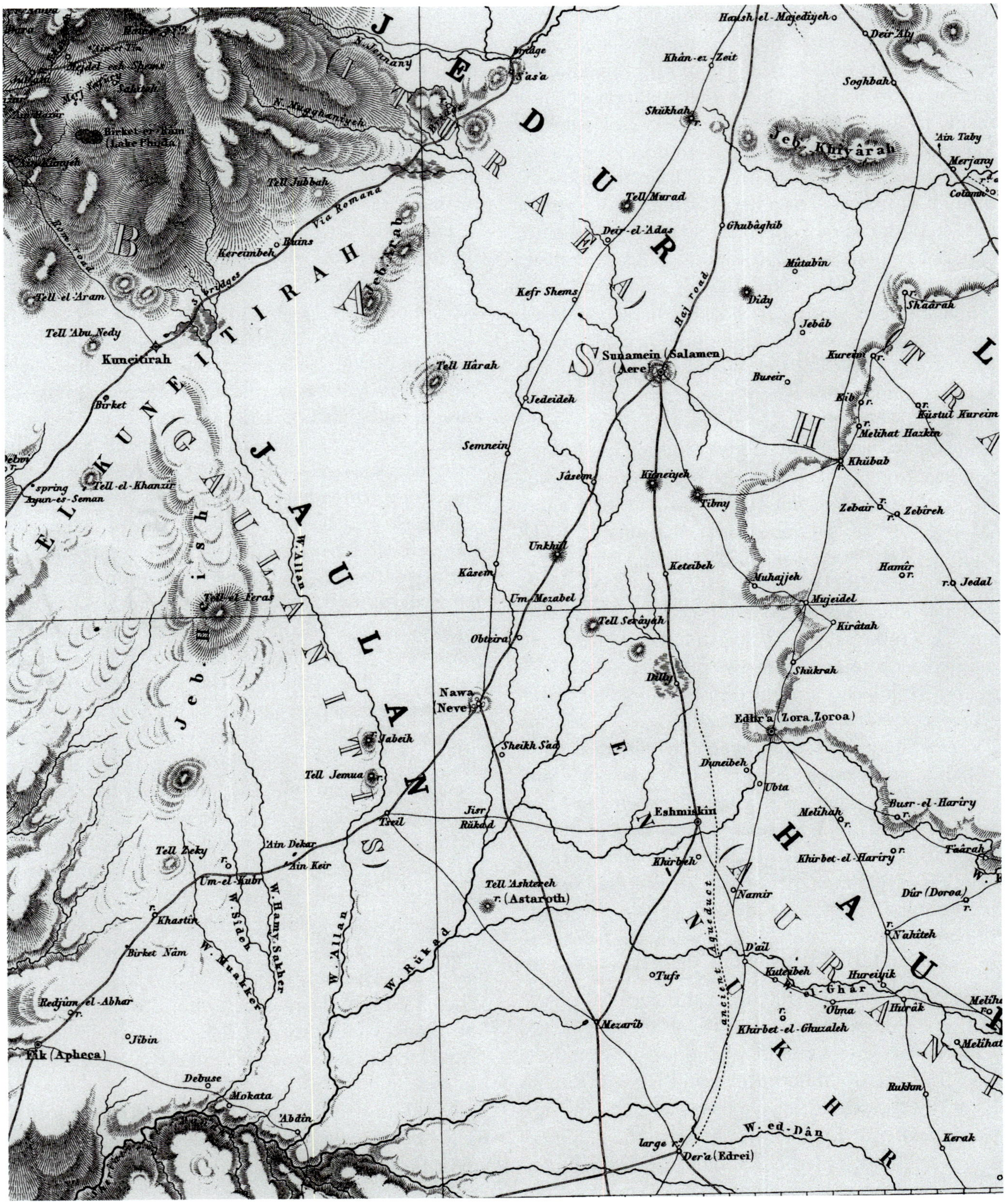

FIGURE 44. C. W. M. van de Velde, from section 4 of 'Map of the Holy Land' (1858).

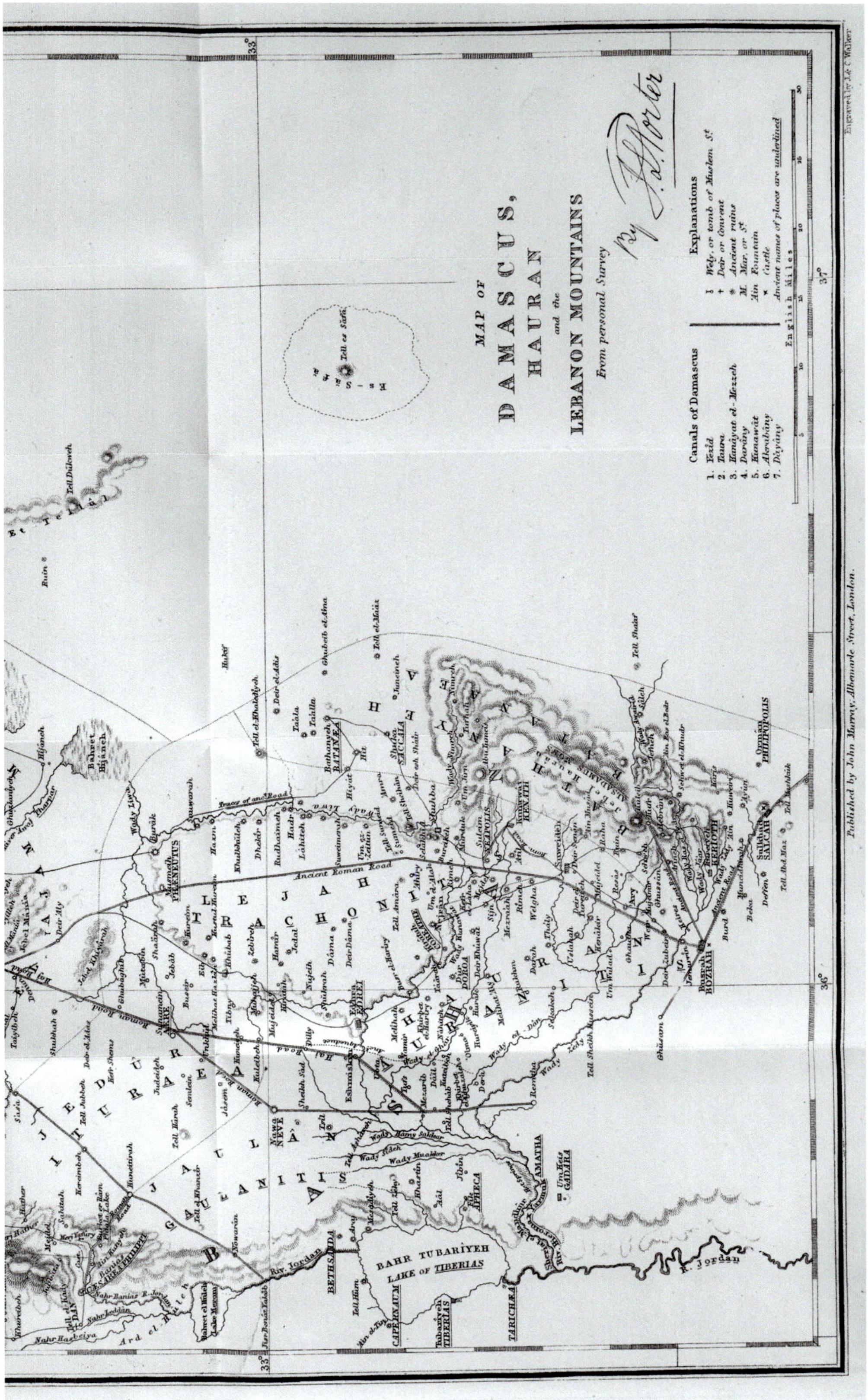

FIGURE 45. J. L. Porter, 'Map of Damascus, Hauran and the Lebanon Mountains from personal survey' (*Five Years in Damascus*, 2nd ed. 1870 [as in 1st ed. of 1855, the date '1855' having been removed]). (*The Board of Trinity College, Dublin*)

FIGURE 46. J. S. Buckingham, map showing his route east of the Jordan (*Travels in Palestine*, 1821). (*The Board of Trinity College, Dublin*)

Some 30 previously known names have been applied to wadis, springs, and other topographical features. About 240 names are already found on the maps of Seetzen, Burckhardt, Buckingham, and Kiepert. For the region between Damascus and the Sheriat el-Mandhur, van de Velde was primarily and directly dependent on the map of J. L. Porter (Fig. 45), entitled 'Map of Damascus, Hauran and the Lebanon Mountains from personal survey 1855 by J. L. Porter', published in Porter's book *Five Years in Damascus: including an account of the history, topography and antiquities of that city, with travels and researches in Palmyra, Lebanon and the Hauran* (2 vols, London, 1855). (Van de Velde notes in his memoir [p. 20], however, that Porter's map had been seriously distorted by the application of the compass variation as if it had been about 10 degrees east instead west.) This map gives, for the regions of Gaulanitis, Hauranitis, Ituraea,

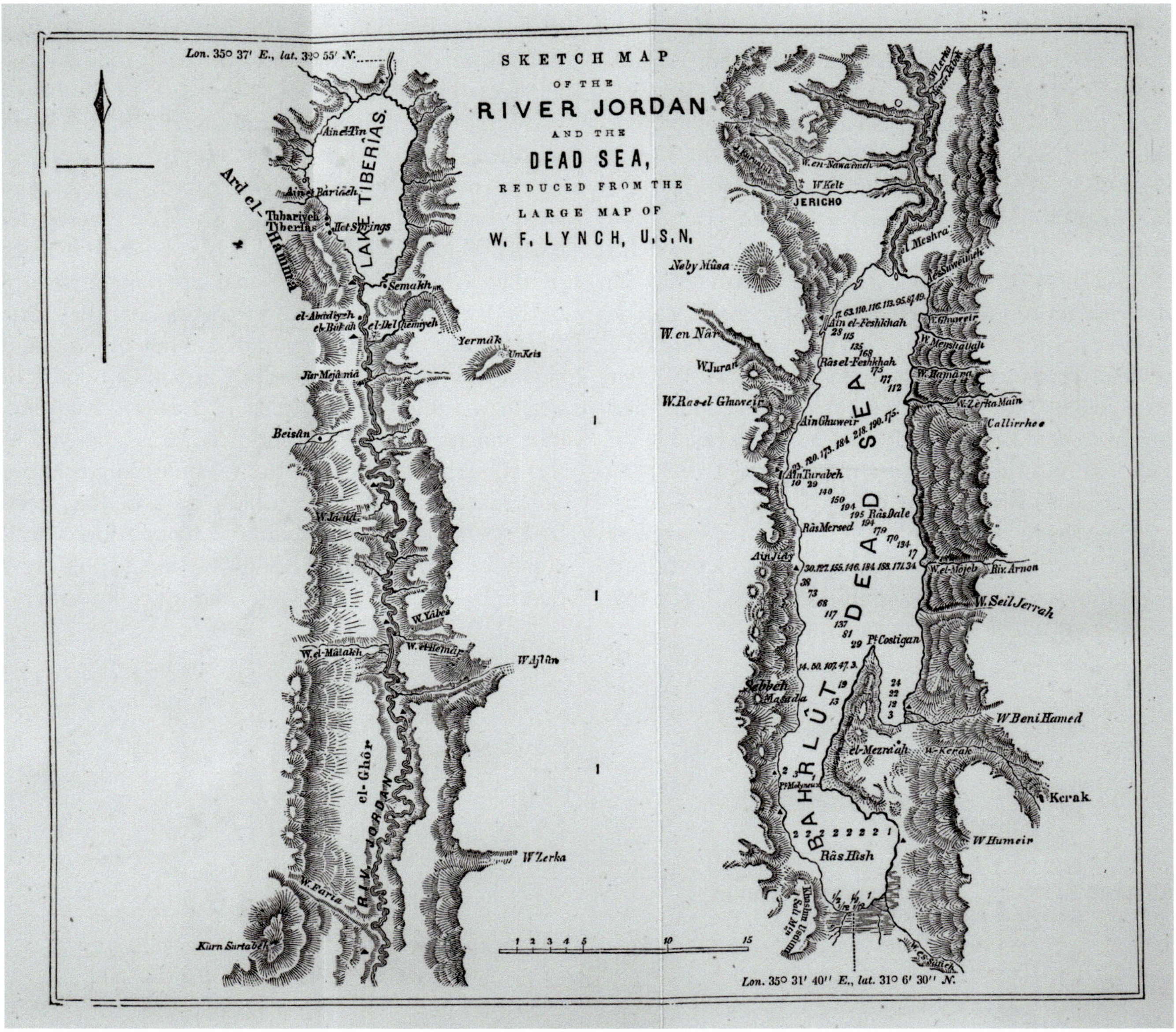

FIGURE 47. W. F. Lynch, Sketch Map of the River Jordan and the Dead Sea, reduced from the large map of W. F. Lynch, U.S.N. (1848) (Lynch 1849). (*The Board of Trinity College, Dublin*)

Batanaea and Trachonitis, about 175 place-names, almost every one of which appears on van de Velde's map. The map from Buckingham's *Travels in Palestine* (London, 1821) (Fig. 46) contributed a small number of place-names from the 'Ajlun region, south of the area later covered by Porter, but at least half of these were known also to Seetzen and Burckhardt. Van de Velde seems to have ignored other of Buckingham's place-names from the 'Ajlun (e.g., Shahan, Ullan, Deer el-Ramjah, Jehaz/Jejaz, Nahr el Zebeen, Samoon, Tugbool, Cufr Sow, Jejean), perhaps because they were unsupported by other travellers. For the Transjordanian place-names south of the River Jarmuk, Van de Velde relies mainly on the work of Seetzen, Burckhardt and Kiepert. For the Dead Sea, Van de Velde draws on the sketch map of Lt W. F. Lynch published in 1848 (Fig. 47), but his outline of the Dead Sea is not exactly the same as Lynch's, and, though he copies Lynch's zig-zag line of depths across the Dead Sea, he does not reproduce Lynch's figures precisely. Van de Velde revisited Palestine in 1861–62, and published a revised version of his map in 1866. (See Fischer 1940, 65–68, and Ben Arieh 1979, 147–52.) His map has been generally regarded as the best available for Palestine until the publication of the *Map of Western Palestine in 26 sheets* in 1880.

The eighteenth-century gentlemen-travellers found worthy successors in the more determined explorers of the nineteenth century. Men like Seetzen, Burckhardt, Irby and Mangles and others like Leon Laborde and the Comte de Bertou travelled purposefully and recorded what they found. Robinson and Smith researched in scholarly manner the relationship of the ancient biblical toponyms to the modern Arabic names. Artists like W. H. Bartlett and David Roberts travelled the land, picturing it in romantic fashion to meet a curious market in Britain and America. Cartographers like Kiepert and van de Velde up-dated the maps, applying the science of compass-bearings to the problems of the cartography of Transjordan. Van de Velde could take his starting point from points fixed west of the Jordan by the triangulation of Symonds. The work of these two cartographers, though a great advance on the work of their predecessors, was almost immediately rendered obsolete by the work of the professional surveyors of the British Army's Royal Engineers and others. Van de Velde's map was published in 1858; in 1867 Charles Warren began his reconnaissance of the Jordan valley, in 1872 the American Palestine Exploration Society began its survey of the land east of the Jordan, and in 1881 Captain Conder, RE, began a triangulated survey of Eastern Palestine. The northern part of Transjordan was surveyed in the 1880s and 1890s by the German engineer G. Schumacher, and the southern part in the 1900s by Czech scholar and military officer Alois Musil; complete surveys of the region were finally published by the British War Office after the 1914–18 war.

CHAPTER TEN
THE TRIANGULATION OF TRANSJORDAN

CHARLES WARREN

PROFESSIONAL SURVEYING of Transjordan based on triangulation began with Warren's July 1867 'Reconnaissance of the Jordan Valley'. Warren seems to have covered the whole Transjordan region from Jisr el-Mejamia in the north, to Amman in the east, and to Callirhoe in the south, including the Zerka basin, the regions of Amman, Es-Salt, Arak el-Emir, Hesban and the Belka. His work involved 'the taking of a great many photographs of the old cities of Rabbath Ammon, Jerash, Ramoth Gilead, and others; observations of latitude and longitude in fourteen places; and a long list of 350 names, of which 84 agree nearly with 84 out of the 135 taken by Dr Robinson' (Committee of the PEF, 1873, 221–22). The archives of the Palestine Exploration Fund contain two manuscript sketch maps of Warren's 1867 reconnaissance; the first (PEF/JER/WAR/66) is described as 'Sketch map of the area (mainly) east of Jordan. [n.d. ?1867]. Scale approx. 1 inch: 2 miles. 67 × 50 cm.'. This appears to be Warren's original sketch map. The second (PEF/JER/WAR/67) is headed 'Reconnaissance of the country between the two Zerkas and the Jordan, 18th July–15th August 1867. Scale 1 inch: 2 miles. 68 × 78 cm., and signed 'Charles Warren, Lieut. RE', and dated '2nd Nov. 1868'. This appears to be Warren's final hand version of his reconnaissance map, and carries a lengthy note in Warren's hand giving details about heights, levels, latitude and longitude, and the transcription of Arabic place-names. This map (scale 3 miles to one inch) (Fig. 48) was not published until 1880, in *Excavations at Jerusalem 1867–70*, plate XLVIII (though no. 49 on the Contents page).

THE MAPS OF E. H. PALMER AND H. B. TRISTRAM

South and east of the region covered by Warren's reconnaissance lay the land of biblical Moab. The discovery of the Moabite Stone in August 1868 by a CMS missionary, the Revd F. A. Klein, led to new interest in the region of Moab. In 1868–69 a Palestine Exploration Fund team led by Captains Charles W. Wilson and H. S. Palmer, RE, accompanied by the Arabist E. H. Palmer and others, surveyed the Sinai region. In late 1869 E. H. Palmer, commissioned by the PEF to explore the desert of et-Tih, accompanied by C. F. Tyrwhitt-Drake, commissioned by the University of Cambridge to report on its natural history, set off from Suez towards the desert of et-Tih and the Negev. In April they turned their attention to the Wadi 'Araba, Petra, and the region of Edom; by May they were in the land of Moab, passing by the Ghor es-Safi, the Lisan, the Wadi Kerak, the Wadi Mojîb, Dibhân, Um Rasas, the Wadi Waleh and Mt Nebo before reaching the Jordan. Palmer described the journey in *The Desert of the Exodus* (Cambridge: Deighton Bell & Co, 1871), and published in it, 'by permission of the Committee of the Palestine Exploration Fund' (I, viii), a coloured 'Map of Moab embodying the investigations of Captain Warren R.E. and E. H. Palmer Esq' (facing p. 471) (Pl. 11). This presumably refers to Warren's reconnaissance of July 1867; Palmer seems to have had access to Warren's sketch map or to the early copy of it, and used it for the Belka and Hesban regions, where his journey ran through territory reconnoitred by Warren.

In 1872, the scholar H. B. Tristram, 'stimulated' by a grant from the British Association 'for the purpose of undertaking a *Geographical* exploration of the

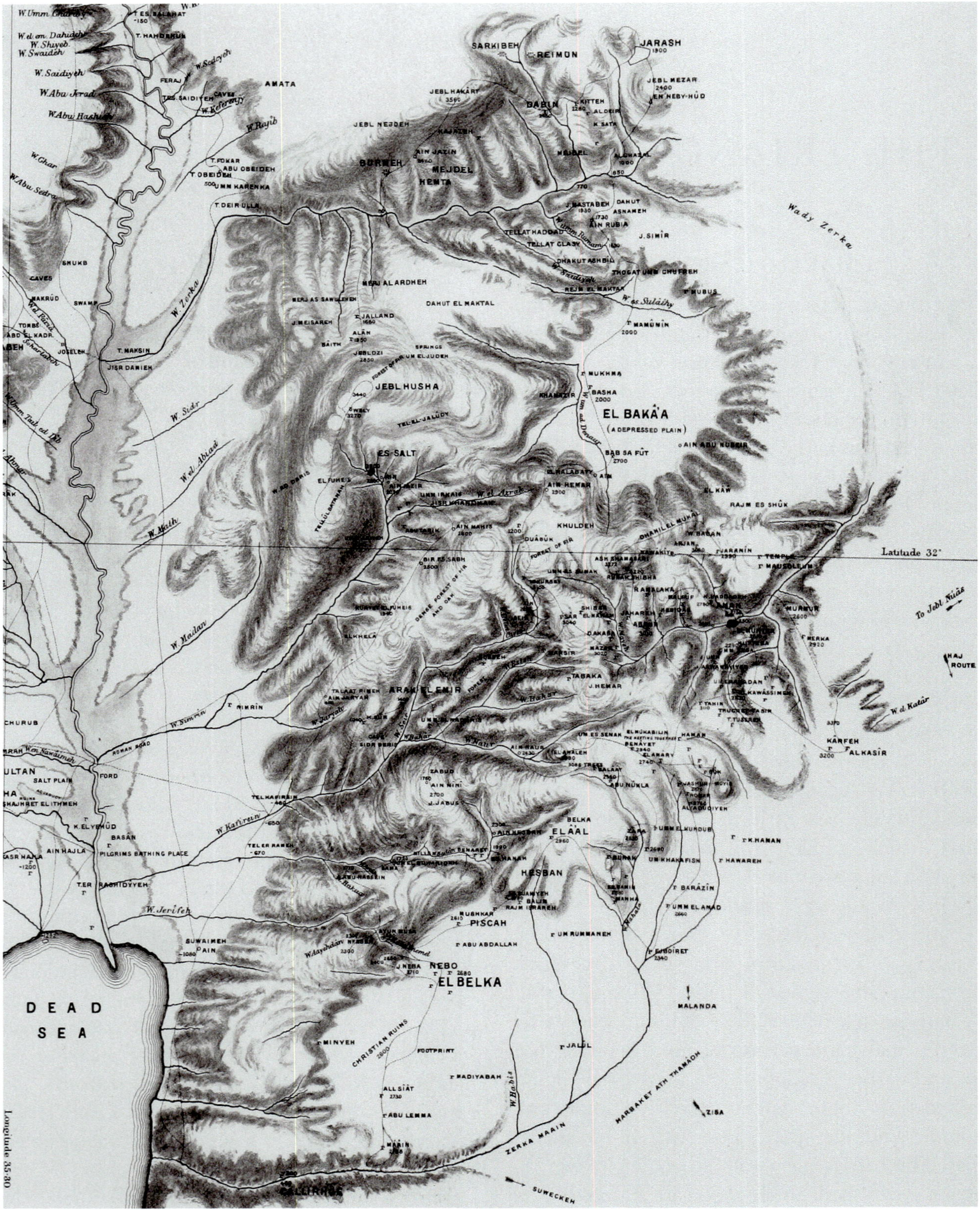

FIGURE 48. Charles Warren, 'Reconnaissance of the Jordan Valley' (1867; pub. 1880).
(*The Board of Trinity College, Dublin*)

country of Moab' (Tristram, 1873, 3), also led an expedition to Moab. Tristram was critical of Palmer's journey (as the italicised *Geographical* implies), noting that it 'was comparatively barren of results, though it added something to our topographical knowledge', and expressing the hope that 'a careful survey might probably add much to our topographical knowledge, and at least decide the position of many ancient sites, marked at random on the present maps' (ibid.). Tristram also approached Moab via the southern end of the Dead Sea, and ascended to Kerak. After being held hostage for a short time, the party proceeded north to Rabba, past Jebel Shihân and across the Wadi Mojib, and past Arair to Dhibân. From here they made an expedition to Um Rasâs and then north to Um el-Walid and Qasr Mushatta before returning westwards towards the Zerka Maîn, Callirhoe, Machaerus, and then back towards Medaba, Nebo, and Heshbân, and the Jordan. Tristram also published a coloured 'Map of Moab' (Pl. 12) showing his route (facing p. 1). It is clear from the maps and the routes given that Tristram saw much more of the country than Palmer did. Tristram's map has about 160 place-names, Palmer about 120; about 60 are common to both. Tristram adds a large group of names unknown to Palmer in the region of Um Rasâs east of the W. Mojîb, to which Palmer paid only a flying visit ('a bold dash eastwards', according to Tristram), and again in the region of Kerak, which Palmer did not visit at all.

The two maps of Palmer and Tristram (1871 and 1873) are in some ways very similar. They are based on the same coordinates, and the outline of the Dead Sea is almost identical. The wadi courses given are similar, but not identical; in particular, Palmer connects his Wadi Themed with the Zerka Main, and Tristram with the Seil Heidan and the Wadi Mojib. The modern map suggests that Tristram may be the more correct. On the other hand, Palmer's presentation of tributaries to the Mojib from the south seems more accurate than Tristram's. Tristram places Um Rasas more precisely than Palmer, who has it too far north-east. Palmer's Ataroth (Attarûs) is correctly placed further north than Tristram's. Both maps show er-Rîha well south of Jebel Shihân, on which point see Miller above. These two maps demonstrate how urgently a professionally surveyed map of the region was needed.

THE SURVEY OF THE AMERICAN PALESTINE EXPLORATION SOCIETY

In 1872 the American Palestine Exploration Society, founded in 1870, embarked upon a survey of the land east of the Jordan. The intention was to publish it alongside the Palestine Exploration Fund's *Survey of Western Palestine*. The actual work of exploring and surveying took place in 1873, and 1875–77. The territory covered reached from just north of Banias to just south of Madeba; the upper reaches of the Yarmuk basin were explored as far east as Salkhad. By 1880, however, it was clear to the Palestine Exploration Fund in London that the American Exploration Society's maps were not sufficiently accurate in detail to be related to the Fund's large-scale map — there were, for example, serious discrepancies between the two maps in the course of the Jordan — and that the Americans had produced a reconnaissance rather than an exact survey. A map of thirteen sheets covering some 5,700 square miles was printed by the British Ordnance Survey in Southampton in 1879, but this map, though of interest to the Intelligence Department of the War Office in London, was not published (for the Index map, see Fig. 49), and the project of a joint American-British survey covering both sides of the Jordan lapsed. (For a detailed account, based on the archives of the Palestine Exploration Fund, see Cobbing 2005.)

THE PALESTINE EXPLORATION FUND'S *SURVEY OF EASTERN PALESTINE*

It was left to the British to complete the survey of Transjordan, and in 1881 Captain Conder, RE, and Lt Mantell, RE, initiated a triangulated survey of Eastern Palestine, beginning with the southern half. Their survey in July-November 1881 extended approximately from Wadi Nimrin in the north to Wadi Zerka Main in the south, reaching Amman in the east, some 510 square miles in all. However, the opposition of the Turkish governor of Es-Salt prevented any further progress, and on 15 November 1881 Conder was formally ordered to withdraw from

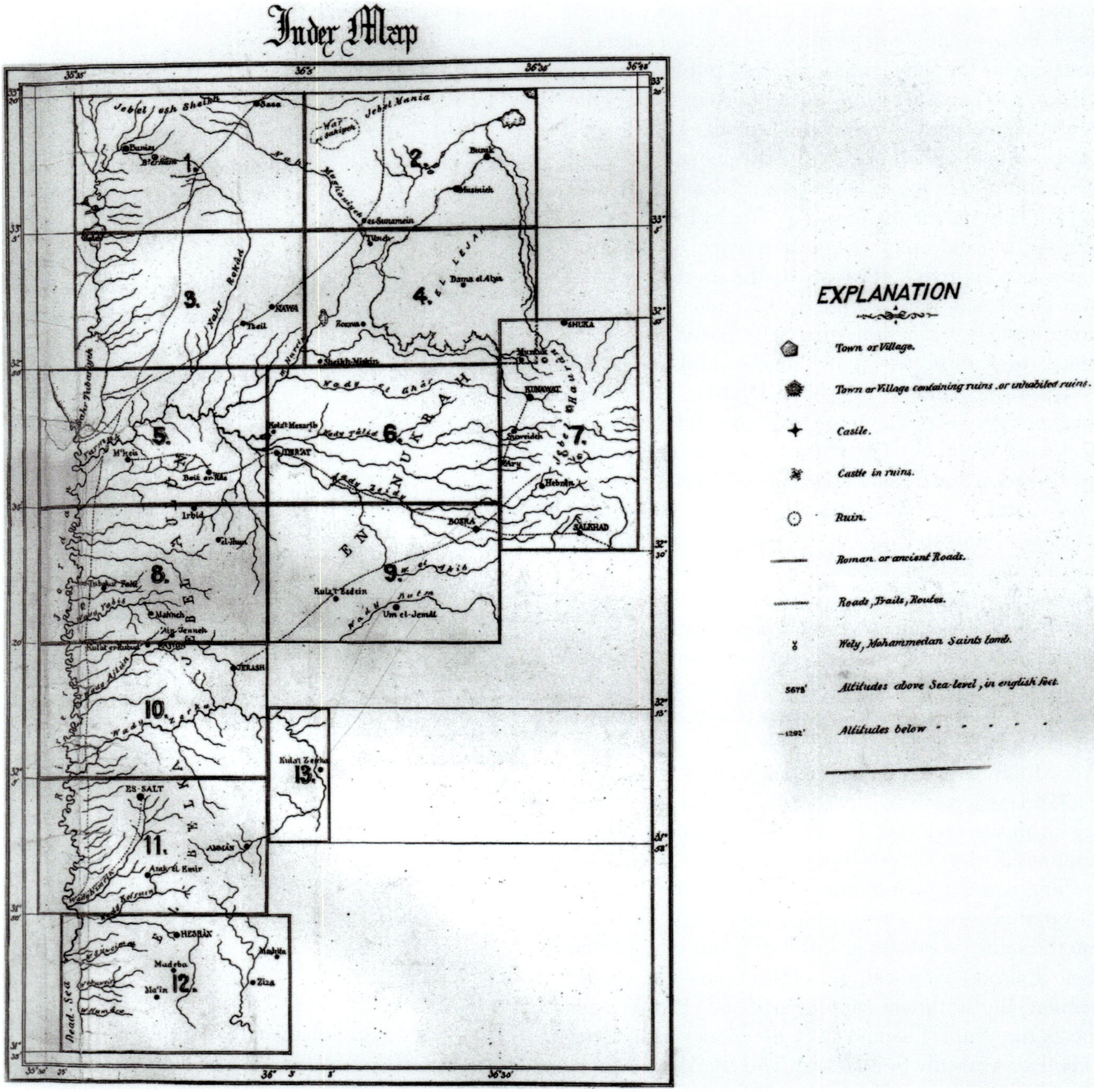

FIGURE 49. Index map of the American Palestine Exploration Society's Survey of Eastern Palestine 1879 (Cobbing 2005). (*Palestine Exploration Fund, London*)

the work (see Moscrop 2000, 129–36). Conder's volume The Survey of Eastern Palestine: Memoirs of the topography, orography, hydrography, archaeology etc, Vol. 1: The 'Adwan country (London, PEF, 1889), contains his map of the region surveyed (Pl. 13), and a diagram of the triangulation, showing its connections with the western survey (ibid., p. 287) (Fig. 50). Survey of the rest of Transjordan was to follow piecemeal, the work of a number of scholars and surveyors, over the following decades.

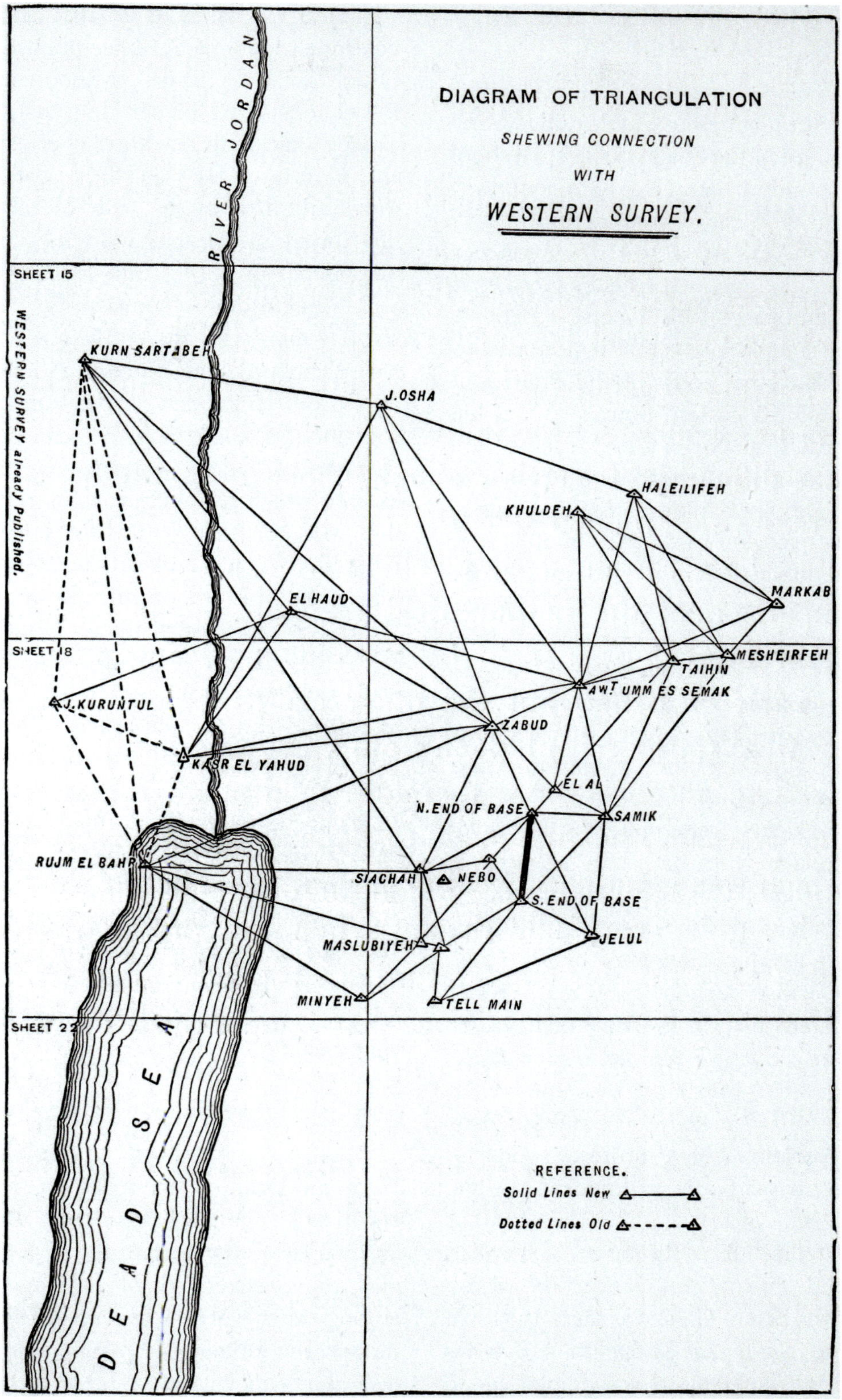

FIGURE 50. Diagram of triangulation of Conder and Mantell's survey (1881), showing connection with Western Survey (1872–80). (*Palestine Exploration Fund, London*)

THE SURVEYS OF GOTTLIEB SCHUMACHER

In the 1880s Gottlieb Schumacher, a German engineer working for the Damascus-Haifa railway, surveyed the regions east of the Sea of Galilee. His book *Across the Jordan* (London: Bentley, 1886) contains a map entitled 'Map of Western Haurân and Eastern Jaulân from the Surveys of Gottlieb Schumacher, C.E., reduced from the one-inch scale by Geo. Armstrong for the Committee of the Palestine Exploration Fund'. This map covers the region east of the Nahr er-Rukkād and north of the Sharîat el-Menâdire (Yarmuk); its eastern boundary is a line running north-south from Nawâ to el-Mezeirib. The Preface to this book by W[alter] B[esant], Honorary Secretary to the Palestine Exploration Fund, suggests that the work of surveying took place in 1885, and that the book is a publication of the Palestine Exploration Fund. Schumacher's second book, *The Jaulân, surveyed for the German Society for the Exploration of the Holy Land* (London: Bentley, 1888; translated, by permission, from the Transactions of the German Society) contained a 'Map of the Jaulân by Gottlieb Schumacher, C.E. 1885', with 'Palestine Exploration Fund' printed across the top right hand border. This map covered the region between the upper Jordan valley to the Wadi er-Rukkād from Banias in the north to the Yarmuk in the south. This map also was reduced from Schumacher's work by George Armstrong, Besant's successor as Hon. Secretary to the Fund. A slightly fuller German version of this map, 'Karte des Dschōlān', dated 1885, in the scale of 1:152,000 was published in *Zeitschrift des deutschen Palästina Vereins*, Bd IX, in 1886. A third book by Schumacher, *Northern Ajlun 'within the Decapolis'*, was published by the Palestine Exploration Fund in format similar to that of the previous two books in London in 1890. This contained a 'Map of a part of the Kada Irbid or Northern 'Ajlûn reduced from the survey of Gottlieb Schumacher, C.E.' (Fig. 51) and covered the region south and east of the Sea of Galilee as far as the town of Tibneh. The preface to *The Jaulân* (pp. 5–8) notes that Schumacher's triangulation was joined on to that of the PEF map of Western Palestine. Clearly Schumacher's work was claimed by the Fund as a natural extension of its own Survey of Western Palestine. Schumacher later published a 'Karte des Ostjordanlandes', scale 1:152,000, in two sheets: Blatt I: 'Dschōlān & westlicher Haurān' (ZDPV 22 [1899], Tafel 4), and Blatt II: 'Nördlicher 'Adschlūn & südlicher Haurān' (ZDPV 20 [1897], Tafel 1). Some sheets were published in the scale of 1:63,360 (matching the Survey of Western Palestine) in 1908 and 1911 (see Fischer 1913, 150).

It is interesting to compare Schumacher's Transjordan maps with the relevant sections of Van de Velde's map, which represents the final generation of the untriangulated maps. Van de Velde's map (1:315,000) is about half the scale of Schumacher's (1:152,000 [ZDPV], ⅜ in. to one mile [PEF]), and extends much further east than Schumacher's. For the more northerly part of the Jaulan, Schumacher takes the W. er Rukkād as the eastern limit of his survey. Most importantly, however, he places the W. er-Rukkād west of the W. 'Allan, whereas van de Velde places the Wadi 'Allan west of the W. er-Rukkād (Rŭkad), consequently misplacing Jisr er-Rukkād east of Tseil instead of west. In the Irbid region, van de Velde sets Irbid east of Beit Rās, instead of south, el Hŭsn where Irbid should be, and Maru north of Beit Rās instead of east; Schumacher places Irbid and Maru correctly, but el Hŭsn lies off his surveyed area to the south. There is no doubt, however, that Schumacher's is the much more accurate map, giving many more local place names, and correctly locating mountains, wadis, springs, bridges, and various ruins as well as towns and villages.

H. FISCHER AND H. GUTHE: *PALÄSTINA 1:700,000*

Hermann Guthe (1849–1936) was a biblical scholar, occupying a chair at Leipzig from 1884, with particular interests in the geography, cartography, topography and archaeology of Palestine. He published among other works a history of Israel (1899) which showed the influence of Wellhausen's work, a Bible dictionary (1903), and a Bible atlas (1911; 2nd ed. 1926). 'His atlas was one of the first to be based on a critical reading of the biblical text and carefully

FIGURE 51. G. Schumacher, 'Map of a part of the Kada Irbid or Northern 'Ajlûn' (*Northern 'Ajlun 'within the Decapolis'*, 1890).

distinguished site identifications that were certain from those that were not' (Graham 1999, 474). In 1906 he published an edition of the Madeba mosaic map, with coloured plates. He helped establish the learned society of the Deutscher Verein zur Erforschung Palästinas, and edited its journal, largely dedicated to archaeological and related studies (*Zeitschrift des deutschen Palästina-Vereins*) from 1878–96. Hans Fischer (1860–1941) was a cartographer from the Geographical Institute in Leipzig. His first map (1889) was of the Hauran region at 1:400,000 (*ZDPV* 12 [1889], Taf. 6); his second was the map of Palestine published by Guthe in ZDPV in 1890, with a revised edition in 1913. Fischer was subsequently responsible (Alt 1942) for a wall-map of Palestine in biblical history (1896), and for the cartography in Guthe's *Bibelatlas* (1911); the revised edition (1926) was described by the Old Testament scholar Martin Noth as 'currently the best survey map of Palestine; it is the most reliable and at the same time the most detailed' (Noth 1964, 5; English translation of Noth's 1962 revision of a work originally published in 1940).

Fischer and Guthe's 1890 map (Pl. 14), basically dependent on the work of Schumacher and the PEF, was essentially a map of contemporary Palestine, showing its political divisions, the cities, towns and villages by population levels (over 10,000; 3000–10,000; 500–3000; and smaller places), Christian churches and Muslim holy-places, and ruins. Contemporary Arabic names are given, with historical names added in 'Lapidarschrift' (*sans serif*). In Transjordan, comparatively few historical or biblical sites are identified; these include, from north to south, Baneas/Caesarea Philippi, Phiala, Chasfon, Gadara, Edrei, Gerasa, Gilead, Jogbeha, Rabbath Ammon/Philadelphia, Eleale, Hesbon, Medaba, Beth Baal Meon, Callirhoe, Ataroth, Kirjathaim, Rabbat Moab/Areopolis, and Kir Moab. This map extended from north of Beirut to just south of the Ghor at the southern end of the Dead Sea; to the east it included the Jebel Hauran (of which Fischer had published a map in the previous year), but further south its eastern boundary was the pilgrim route from Damascus to Mecca, the south-eastern corner of the map being taken up by a plan of Jerusalem.

R. E. BRÜNNOW AND A. VON DOMASZEWSKI: *UBERSICHTSKARTE DES OSTJORDANLANDES 1:700,000*

Rudolf-Ernst Brünnow (1858–1917) (whose father was from 1865–74 Astronomer Royal in Ireland and Professor of Astronomy at Trinity College Dublin) was a classical scholar who in 1897 and 1898 travelled in southern Transjordan with the ancient historian and archaeologist Alfred von Domaszewski (1856–1927), then a professor in Heidelberg, systematically exploring and recording Roman sites and road systems. The result was *Die Provincia Arabia* (three volumes, published by Trübner in Strassburg, 1904–09), which Wellhausen (1904, 943) described as 'a lasting monument' (*ein dauerndes Monument*). Volume I contained their map, *Ubersichtskarte des Ostjordanlandes* (Pl. 15). The caption describes this map, on the same scale as the Fischer-Guthe map, as 'Ausschnitt aus der Fischer-Guthe'schen Karte, mit Benützung der noch unveröffentlichten Aufnahme des südl. 'Adschlun von Dr G. Schumacher, und im Süden berichtigt und bis Petra erweitet durch die Aufnahmen in 1:100,000 von Prof. Dr R. Brünnow'. Thus this map used a section of the Fischer-Guthe map, with unpublished information from Schumacher's work on the southern Ajlun, and was extended as far south as Petra with information from Brünnow's larger map. George Adam Smith (1904) published a useful critique of *Die Provincia Arabia*, vol. 1 (including the map). He noted that Brünnow and v. Domaszewski

> have carefully surveyed the land of Moab south and east of the part covered by Colonel Conder's survey, and the land of Edom south of that. Professor Brünnow's triangulation is grounded on a great triangle formed by the three far-seen summits of Shihān, Jebel Jiyāl and Hafret Ka'dân, with a measured base of 1,000 metres in the neighbourhood of el-Kastal. He has connected this triangulation with that of the English survey, and claims for it a correctness north of the Wady el-Hesa within 200 metres. South of that wady he was unable, on account of weather, to reach the same degree of correctness, but believes that even here the probable error is not more than 500 metres. These calculations were further tested by determinations of latitude, frequent and careful notes of the times occupied on his journeys between different points and by sketches of variations in the level of the ground. To all this were added numerous observations of heights by aneroid.

The calculations have been reworked by Professor Kunze. (Smith 1904, 297–98)

Smith importantly draws attention

to the way in which Professor Brünnow's results have revolutionised our knowledge of the tributaries of the Môjib and of the wadies which combine to form at el-Kerak the Wady el-Kerak. Hitherto the latter have been supposed to rise far to the east of el-Kerak, near the Hajj road. But Professor Brünnow has proved the existence of a high ridge immediately to the east of Kerak, and the rise of the Kerak wadies to the west of this and south of el-Kerak; and shown that the northern tributaries of the Môjib extend east of el-Katraneh and the Hajj road, and far south over the plateau to below the latitude of Kerak. (Smith 1904, 400)

It is thus clear that, as the caption on Brünnow and von Domaszewski's map indicates, this map includes a more accurate survey than the Fischer-Guthe map of the land south of the region surveyed by Conder. The Fischer-Guthe map does indeed suggest that the upper tributaries of the Wadi Kerak rise to the east, which means that this region had not yet been fully surveyed. Examination of Brünnow and von Domaszewski's map and comparison with the Fischer-Guthe map show that the former contains more detail than the latter, including more place-names and the line of the new Turkish railway from Damascus south towards Mecca. Smith noted in his review that 'a very great deal remains to be done before we have a map of southern Moab as reliable as that which we possess for the northern part' (Smith 1904, 399–400). This work was already being done by Alois Musil.

ALOIS MUSIL'S *KARTE VON ARABIA PETRAEA*

The Czech orientalist Alois Musil (1868–1944) travelled in the Transjordan in 1896–98, exploring the desert castles, including Kuseir 'Amra, and collecting material for a 1:50,000 map of the Madeba region which he produced in 1898. His interest further aroused by Brünnow's work, he travelled again in 1901–02 and drafted a 1:50,000 map for the whole of Arabia Petraea. His map of Arabia Petraea ('Karte von Arabia Petraea nach eigenen Aufnahmen von Professor Dr. Alois Musil', at a scale of 1:300,000) (Pl. 16, Fig. 52) was published in three sheets in his *Kusejr 'Amra* (Wien: Kaiserliche Akademie des Wissenschaften, 1907) and also in his three-volume *Arabia Petraea* (Wien, 1906–08). (In the preparation of this map, Musil worked as an army officer seconded to the Kaiserliche Akademie des Wissenschaften, assisted by professional surveyors and draughtsmen from the army; cf. *Kusejr 'Amra*, 1, vii.) These three sheets covered between them the area between 34 and 37 degrees east of Greenwich, and 29 degrees 20 minutes and 32 degrees north; in the Transjordan this meant the area from just north of Hesban to the northern end of the Gulf of 'Aqaba, and eastwards beyond the new Turkish railway from Damascus into Arabia. Conder's work in 1881, published in 1889, had mapped the ground from the W. Nimrin to the W. Zerka Main; Musil's work overlapped as far as the Zerka Main, but extended south into the W. Hismeh and beyond, providing the first accurate map of the southern Transjordan for historians, politicians and soldiers alike. A comparison of the work of Conder and Musil, where they overlap east of the northern end of the Dead Sea, shows close similarities. For example, the same wadis appear along the Dead Sea coast, and the same place names appear along the road between Hesban and Madeba. (Fischer notes [1913, 146] that 'die Ostküste des Toten Meeres nördlich vom *rōr el-mazra'* hat Musil uberhaupt nicht berührt'.) Musil's scale is smaller (1:300,000) than Conder's (about 2½ miles to the inch), but his map is clearer, with more detail, and shows the newly built Hedjaz railway line (work began in 1901 and the line from Damascus to Tebuk was complete by September 1906) and telegraph stations. It also shows coordinates, which Conder's unfinished map does not. Miller (1991, 16) notes of Musil's map that 'although it included some distortions, [it] indicated the approximate locations of well over a hundred ruins between Wadi el-Mujib and Wadi el-Hasa'. Altogether, Musil's map was a very useful and professional map, of considerable archaeological and military value.

TWENTIETH-CENTURY MILITARY AND AIR SURVEYS OF TRANSJORDAN

It is clear that political and military concerns lay behind all the surveys of the Transjordan in the late

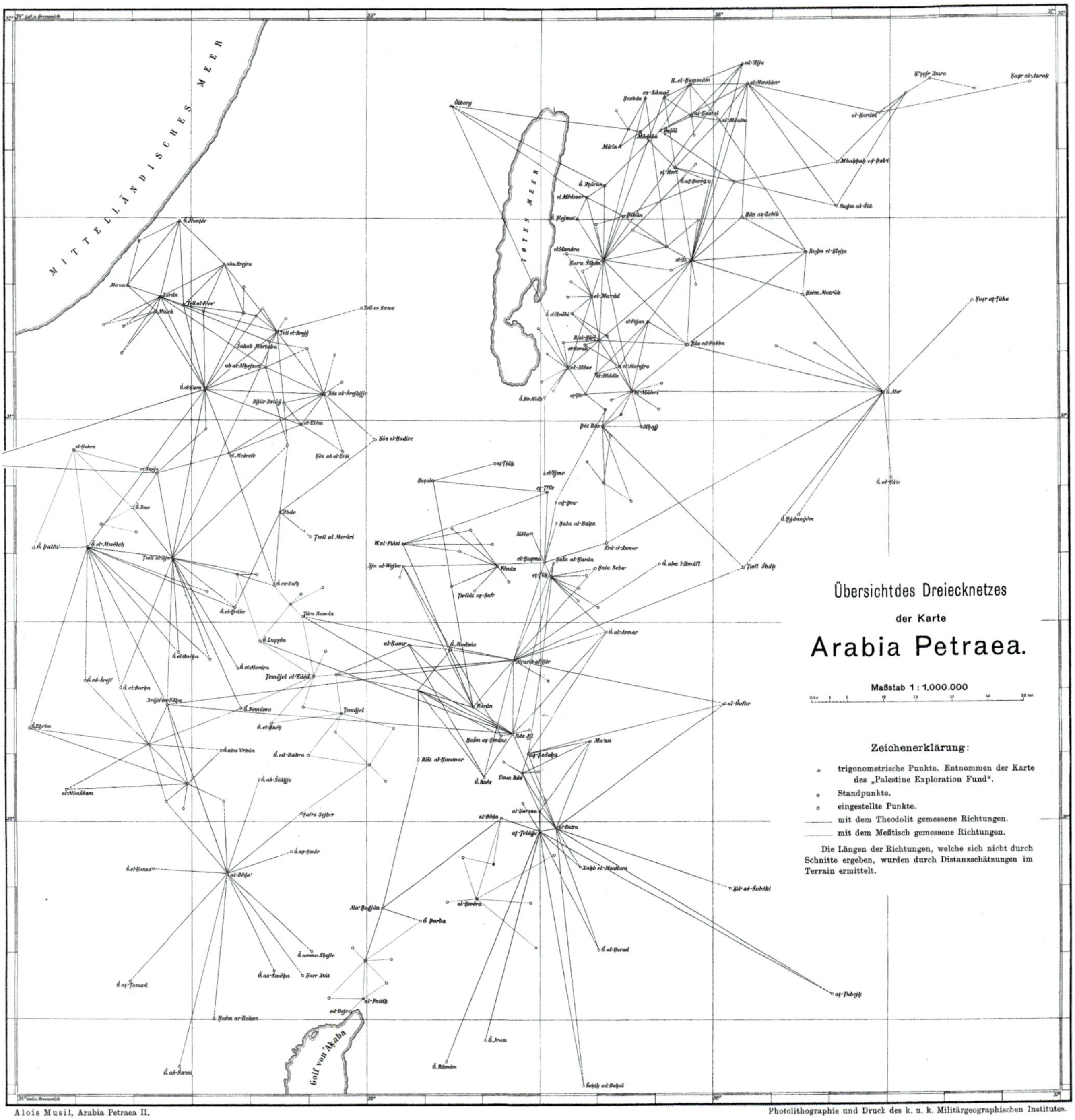

FIGURE 52. Triangulation for Musil's map of Arabia Petraea (Musil, *Arabia Petraea* II, 1908). (*Palestine Exploration Fund, London*)

nineteenth and early twentieth centuries. The Sinai survey of 1868–69, the Survey of Western Palestine in 1872–77, the attempted survey of eastern Palestine in 1881, the survey of the Wadi ‘Araba in 1883, and the survey of the Negev in 1913–14 (which included the whole of the Wadi ‘Araba, and land east of it to a line running north-south from El-Safieh to Petra) were all staffed by officers of the British Royal Engineers, and Musil's work on Arabia Petraea for the Kaiserliche Akademie des Wissenschaft in Vienna was also staffed by military personnel.

Comparison of the ground covered by the nineteenth-century surveys of Transjordan shows that there remained some gaps, at least in the British surveys. Fischer (1913, 148) notes that ‘Der nördlichste Teil der *Belkā*, von den nördlichsten Punkten des englischen Aufnahme bis zum *nahr-ez-zerkā*, war in allen Karten bis jetzt [1913] ausserordentlich mangelhaft dargestellt’. Schumacher's work reached only as far south as Irbid, with a spur south to Tibneh, while the work of Conder and Mantell (1881) mapped the region approximately from Wadi Nimrin east to Amman and south to the Wadi Zerka Main. Warren's 1867 ‘reconnaissance’ covered some of the intervening land — the valley of the Wadi Zerka and the regions of es-Salt and the Beka'a — but not the land between Tibneh and Jerash. This land was covered by the APES survey (whose work was not regarded by the PEF as sufficiently precise to be compared with their own) (Cobbing 2005). From Wadi Nimrin to the Zerka Main Alois Musil's work overlapped Conder and Mantell's, but south of the Wadi Zerka Main Musil's survey reigned supreme.

Newcombe's map ‘The Negeb or Desert south of Beersheba’ (surveyed by, and under the direction of, Captain S. F. Newcombe, RE, 1913–14, and printed for the Palestine Exploration Fund by the G.S., G.S., War Office, 1921), remained unpublished for military reasons until 1921. In 1918 the Survey of Egypt reproduced a set of 1 inch to 1 mile maps of Transjordanian districts (based on the PEF maps) (Fig. 53), prepared by the 7th Field Survey Coy of the RE; the Composite Map of East Jordan (Es-Salt), 3rd ed., dated 15 May 1918, incorporates a note: ‘This map is compiled mainly from Armstrong, Wilson and Conder's 3/8" to mile map of Palestine, except the River Jordan, Wadi Sheib and its tributaries, a small area around Na'aur, and the portion north of wadis Sheib and Ezrak, which have been put in from RAF photographs.’ This marked a new beginning. From now on, aerial surveying played an important role in the mapping of Transjordan. Thus a 21-sheet German military map (1:50,000) of 1918 drew on the work of the PEF, Schumacher, Musil, Fischer and Guthe, ‘ergänzt nach Luftbildern der Feldflieger-Abt’; in 1932–33 the War Office published a set of maps of Transjordan to the scale of 1:50,000 (G.S., G.S., 3937), ‘photographed from the air by no. 14 squadron R.A.F. 1931’; by 1933 16 sheets had been prepared but only sheets 7, 9, 10, 11, 13, and 14 were published. These maps showed contours in red. In 1933 the War Office also published an ‘Air Survey of Eastern Approaches to the Jordan Valley’, on a scale of 1:50,000 (G.S., G.S., 3939) (BMC 48870 [13]), ‘surveyed at the War Office 1931 from air photographs by No 14 (B) Squadron RAF. Ground control under the direction of Major R. Brown R.E. Drawn at the War Office 1932 and printed 1933. Heliographed at the Ordnance Survey 1933’. This pair of maps showed mainly the ‘Ajlun region between the Yarmuk and the Zerka, thus filling the gap between Schumacher's and Warren's work. These maps showed roads, railways, district boundaries, spot heights, telegraph and telephone locations, and police posts (but showed no contours). In 1930 the War Office in London compiled and drew a 1:250,000 map of Palestine and Transjordan with three sheets, from north to south, ‘Haifa’, ‘Jerusalem’, and ‘Maan’ (G.S., G.S., 3903) (BMC 48840 [70, 71]); this was published in 1931 and revised in 1932. In 1933 the War Office published a 1:500,000 map of Transjordan (BMC 48840 [73]). In 1933 the War Office also published the Transjordan at a scale of 1:250,000 in 3 sheets (‘Amman, Ma‘an and Rutba) (G.S., G.S., 3932). In 1937 and 1939 there appeared from the Department of Lands and Surveys, Amman, reproduced by the Survey of Palestine, a map of Transjordan, scale 1:250,000, in three sheets, Amman, Kerak and Ma‘an. After the establishment of the Hashemite Kingdom of the Jordan, these maps were

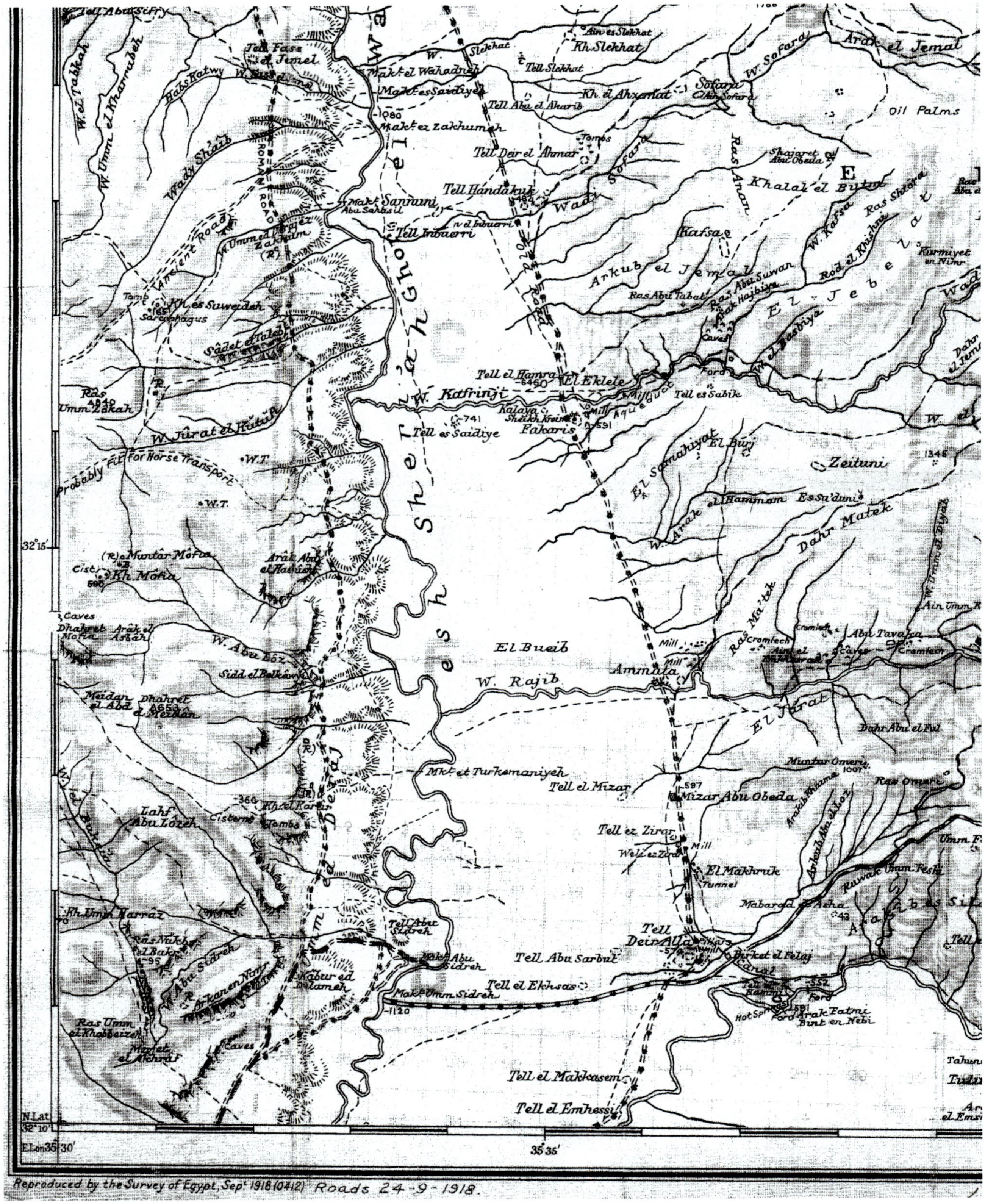

FIGURE 53. Survey of Egypt 1 inch to 1 mile, 1918.

used as the basis for a new set of Archaeological Maps on the sale of 1:250,000 of the districts of Amman, Kerak, Maan and Mahattat al-Jaffur; these were revised and printed by J. N. G. C. and the Directorate of Military Survey, with the cooperation of the Department of Antiquities. These maps from the 1930s and their successors were the cartographical fruit of the British Mandate, and they completed the work of the trigonometric surveys conducted by the Royal Engineers of the British Army in the nineteenth century. They made use of the techniques of mapping by aerial photography, unknown to Warren and Conder, which were in turn superseded in the late twentieth century by the development of satellite imagery. In the 1960s a new set of 1:50,000 maps of Transjordan compiled by photogrammetric methods from aerial photography (Series K737) was prepared for the United States Agency for International Development to Jordan (cf. North 1979, 143–46); this series used the Transverse Mercator Projection rather than the Palestine Grid of earlier British-based maps. The K737 series was used for Miller's Survey of Moab (Miller 1991). An important companion to these maps is the *Gazetteer of Jordan* (2nd ed., 1990), published by the Defense Mapping Agency, Washington D.C.

CHAPTER ELEVEN

THE MODERN IDENTIFICATION OF ANCIENT SITES

One important result of this cartographic activity was that historians of Transjordan, especially those concerned with biblical and classical history, turned their attention increasingly to the identification on the ground of sites named in the ancient sources. By the end of the nineteenth century, as can be seen from the Fischer-Guthe map, not many of these sites were firmly located. In 1907 Peter Thomsen published a valuable list of sites named in literature from the first to the sixth century AD, with a hand-drawn map based on the Fischer-Guthe 1:700,000 map (Thomsen 1907). More identifications were made in George Adam Smith's *Atlas of the Historical Geography of the Holy Land* (1915; 2nd ed. entitled *Historical Atlas of the Holy Land*, 1936). After the Great War, the post-war political reallocation of the territory between the French and the British mandates, and the consequent relative accessibility of the land to academic travellers and archaeologists, a procession of scholars, maps and ancient texts in hand, began to explore the land, seeking to match topography and text. These scholars went in the footsteps of scholarly predecessors, but with much better maps, and, increasingly, with better transport. Between the wars W. F. Albright, Gustaf Dalman, Albrecht Alt, Martin Noth, F.-M. Abel, L.-H. Vincent and others made important contributions to the historical study of topography in Transjordan. M. Avi-Yonah (1925) published an important gazetteer and map of Roman and Byzantine sites (building on the earlier work of P. Thomsen). Nelson Glueck explored the region by car, camel, and on foot, in search of ancient sites of every age, leaving a valuable topographical record in his four volumes of *Explorations in Eastern Palestine* (1933–51) and in his many contributions to the *Bulletin of the American Schools of Oriental Research*, but Glueck did not in fact add many new identifications of ancient sites to the already known list. Martin Noth analysed topographical evidence from the biblical book of Joshua and elsewhere in his commentary on *Joshua* (*Das Buch Josua*, HAT 1.7 [1938]) and in several important monographs. Many of the results of this topographical enquiry are documented in the pages of F.-M. Abel's *Géographie de la Palestine*, II (1938), especially under Part IV, 'Villes bibliques et autres localités historiques', pp. 243–490.

After the Second World War and the occupation of the Jordanian part of the west bank by Israel in 1948, the archaeological exploration of Jordan developed apace. An archaeological map of the Hashemite Kingdom of the Jordan (1:250,000) in three parts — Amman, Kerak and Ma'an — was compiled and drawn by Department of Lands and Surveys of the Jordan, from information supplied by the Department of Antiquities, and published in 1949; the maps themselves were based on the standard 1:250,000 maps, and, clearly drawing on Glueck's explorations, locate most of the sites he visited. These maps were strictly archaeological; they simply marked sites of archaeological interest, giving them their modern Arabic names; they made no attempt to relate them to places mentioned in the ancient sources. The next 50 years witnessed an explosion of activity on the ground in Jordan and in the pages of the journals as archaeologists and historians related the findings of their researches to the topography. A useful summary of sites discovered in the territories of the Iron Age Ammonites, Moabites and Edomites, together with maps and a list of biblical site identifications for Jordan, is given by MacDonald (2000, 157–208, 209–12). MacDonald opens his book with an essay on the history and methodology of site identification (MacDonald 2000, 9–20); an earlier article by J. M. Miller (1983) discussed some of the difficulties in site identification, with particular reference to some Jordanian sites. A. Dearman (1989, 170–96) discussed the identification of places named on the ninth-century BC Mesha Stele. The late Michael Avi-Yonah has bequeathed to topographical scholars a fully-referenced *Gazetteer of Roman Palestine*, complete with maps (1976); this gazetteer includes Transjordanian sites. In 1937 F. J. Salmon, Commissioner for Lands and Surveys in Palestine, had produced, 'in a form derived from Ordnance Survey practice' (Cruickshank 2004, 14) 'Palestine of the Crusades: A map of the country on scale 1:350,000 with historical introduction and gazetteer' (reprinted April 1940, with corrections and additions). A successor to this map entitled 'The Crusader Kingdom of Jerusalem 1099–1291' appeared in 1972 on the slightly smaller scale of 1:400,000, so covering a slightly larger area from north-south but narrower area east-west, thus omitting a castle from Gaza and a church from Amman (Cruickshank 2004, 13). These scholars are following in the steps of Eusebius, scholar and bishop of Caesarea (*c.* 260–340), whose *Onomasticon* of biblical and other toponyms has recently been re-translated, and published together with an annotated index and other essays, and a set of maps (Freeman-Grenville, Chapman and Taylor 2003). Vandenhoeck & Ruprecht of Göttingen published in 1981 a historical-archaeological map of Palestine (*Palästina: historische-archäologische Karte*), with a full index of names both ancient and modern, and with an introduction by E. Höhne; this map includes Transjordan, and marks and identifies sites of all historical periods. These works are indispensable to the modern student of the historical topography of Jordan, and they would have been impossible to produce without the work of the professional cartographers of the late nineteenth and early twentieth century.

HISTORICAL ATLASES

There now exists a large number of historical atlases dedicated to presenting the ancient world — biblical, classical, Byzantine, Islamic, medieval — of Israel/Palestine and Jordan. Classical Transjordan is well covered by the *Barrington Atlas of the Greek and Roman World* (ed. R. J. A. Talbert 2000). Many atlases relate specifically to the biblical world; among those with serious academic intentions may be mentioned *The Westminster Historical Atlas of the Bible* (Wright and Filson 1946), *Atlas of the Bible* (Grollenberg 1956), the *Oxford Bible Atlas* (May, Hunt and Hamilton 1962, 3rd ed., revised Day 1984), *The Modern Bible Atlas* (Aharoni and Avi Yonah 1979), and *The New Atlas of the Bible* (Rogerson 1985). With these must be listed *The Geographical and Topographical Texts of the Old Testament* (Simons 1959). and *The Land of the Bible* (Aharoni 1966, 2nd ed. 1979), which between them supply a detailed study of the historical geography and topography of the Hebrew Bible, including the land of modern Jordan. George Adam Smith's *Atlas of the Historical Geography of the Holy Land* (1915, revised 1936) includes maps of Palestine, west and east of the Jordan, in the fourth century and in the time of the Crusades and the Latin Kingdom of Jerusalem. A much more comprehensive historical atlas of the Levant region, including modern Jordan, is the *Tübinger Atlas des Vorderen Orients* (TAVO), which provides detailed maps of geology, geomorphology, rainfall and hydrogeology, vegetation, animal life, human population, religions, languages, ethnic groups, rural and urban settlements, land utilisation, agriculture, nomadism and sedentarisation, oil and mining industries, trade and tourism, and a large sequence of historical maps covering the region from Lower Palaeolithic times to the twentieth century. Maps from TAVO relating to the history of the land east and west of the Jordan in biblical times have been collected into one volume, *Tübinger Bibelatlas: auf der Grundlage des Tübinger Atlas des Vorderen Orients* (Mittmann and Schmitt 2001). These maps are almost certainly the most detailed and up-to-date available. This volume includes one extra map from more recent times (BX 12) on 'Palestine at around 1920', on which are indicated, *inter alia*, settlements of varying sizes, monasteries, Islamic sanctuaries, ruins, Bedouin migratory areas, roads, railways, and water courses both perennial and temporary.

EPILOGUE

Of the making of maps, as of books, there is no foreseeable end. There are now countless maps of Jordan relating to social, economic, agricultural, mineral, geological, political, military, religious, and other matters. This brief survey of the mapping of Jordan through two millennia shows that much of the extant mapping from past centuries derived from religious and political needs. So in the early centuries the organisational demands of empire resulted in the Peutinger map, and the importance of making pilgrimages to biblical sites led to the creation of the Madeba mosaic map. The maps created over the last century show that historical and religious interest and political and military concerns continue to provide serious motivation for the cartography of the region, and that both religious and academic interests have benefited enormously from political and military requirements. In this last century, the precision resulting from triangulation and photogrammetric techniques has been of great value not only to the military, to security forces, politicians, business people, travellers and tourists, but also to archaeologists, biblical scholars, and other students of ancient topography, and has led to the identification and recovery of many ancient sites whose location has been unknown since their economic decay and physical disappearance. This study has shown that the topography of Jordan has been a subject for serious enquiry, at least among the learned, for two millennia, and has over the last 100 years acquired a life of its own.

TABLE 6. Identification of biblical and classical sites in Transjordan from Robinson and Smith (1841) to the present

Biblical (RSV), classical name	*References*	*Robinson & Smith 1841*	*G. A. Smith 1915, 1936**	*F.-M. Abel 1938*	*Simons 1959*	*A. Aharoni 1979*	*B.MacDonald 2000*	*Others*
Abarim, mountains of	*Dt 32.49; Onom.16.24*			*Range from W. Kefrein to W. Hesa*	*Ridge between Dead Sea & tableland*			
Abel-keramim	Jud.11.33; Onom.32.15			Na'ur	in n.belqa	Na'ûr?	Tall al-'Umayrit?	Sahāb (Knauf)
Abel-meholah	Jud.7.22; 1K.4.12; 19.16; Onom.34.20		'Ain Helweh	T. Abu Sifri		T. Abû Sûs	T. Abu Sus	T.Maqlub (Glueck)
Abel-shittim	Num.33.4; Onom.10.28		Kh. el-Keffrein	T. Kefrein		T. el-Hammâm	T. al-Hammam?	
Abila (in Abilene)	Jos.Ant.19.275; 20.138; Ptol.V.14.18; Onom.32.17		Sūk Wadi Barada	T.Abil				
Abila of the Decapolis	Jos.Ant.12.136; Polyb.5.69–70; Pliny NH.V.74. Onom.32.16; Ptol.5.14.18		Abil	T.Abil				Quailibah (Schumacher)
Abrona	Num.33.34			Ain Defiyé		Umm Rashrash		
Adam 'beside Zarethan'	Josh.3.16; 1 K.7.46 (emend.)		Ed-Dāmieh	T. ed.Damiyé	t.dāmijeh	T. ed-Dâmiyeh	T. ad-Damiyya	
Admatha	Not.Dig.Or.34.33							W.'l-itm (Thomsen)
Adraha (= Edrei)	Tab.Peut.; Ptol. V.14.18, 20; Onom.12.13							
Adrou	Ptol.V.16.4							Odruh (Thomsen)
Alema (cf Helam)	1 Macc.5.26		'Almā	'Almā	'alma			
Almon-diblathaim/ Beth-diblathaim	Num.33.46/ Jer.48.22; Mesha 20				deleilat el-gharbiyeh & esh-sherqiyeh	Kh Deleilât esh-Sherkîyeh	Kh.ad-Deleilat al-Gharbiyya & ash-Sharqiyya	
Amathous	Onom.22.24; Madeba mosaic map		'Ammāta	'Amta				
Ar of Moab	Num.21.28; Is.15.1; Dt.2.9,18,29; Onom.10.25; 44.8; 76.11	Rabba		Er-Rabba	'always a region, never a city' = Moab	Kh. el-Medeiyineh?	Dhiban	Kh.et-Bālū' (Miller)
Areopolis. see. Rabbath Moab								

Biblical (RSV), classical name	*References*	*Robinson & Smith 1841*	*G. A. Smith 1915, 1936**	*F.-M. Abel 1938*	*Simons 1959*	*A. Aharoni 1979*	*B.MacDonald 2000*	*Others*
Arindela	Not.Dig.Or.34.44	Ghŭrŭndel		Kh.Garandel (nr Buseira)				Gharandel (Glueck)
Arnon, river	Num.21.13,26; Jud.11.18; Onom.10.15; Not. Dig.Or.27.34,35	W. el-Mojib	W. Mōjīb					
Aroer (1) on edge of valley of Arnon	Dt 2.36; 3.12; 4.48; Josh 12.2; 13.9, 16; 2 S. 24.5; Jer.48.19; Onom.12.5; Mesha 26		El-'Ara'ir	'Arā'ir	arā'ir	'Arâ'ir	Arair	Kh. 'Ara'ir (Dearman)
Aroer (2) e. of Rabbah	Josh.13.25; Jud.11.26, 33; Num.21.26 (LXX); Mesha 26			'Arğān?				Es-Sweiwîna (Glueck)
Ashtaroth	Josh.9.10; 12.4;13.12,31; Onom.12.1		T. 'Ashtūrah	T. 'Astera				
Ashtaroth-karnaim	Josh.12.4; Onom.6.4, 12.14				t. ash'ari			
Ataroth	Num 32.3,24; Mesha 10–12		Attārūs	Kh. 'Atārūs	h. 'atārūs	Kh. 'Attārûs	Kh. 'Atarus	Kh. 'Atarus (Dearman)
Atroth-shophan	Num 32.35; Onom.12.30						Rujm 'Atarus	Jebel (Rujm) 'Attarus (Tristram)
Avith	Gen.36.35; 1 Chr.1.46			Kh.el-Ğitté				
Baal-meon	Num 32.38, 1 Chr 5.8, Ez 25.9, Mesha 9, 30; Onom.44.21		Ma'īn	Mā'īn	mā'īn	Mā'în		Kh. Ma'in (Seetzen); Ma'in (Dearman)
= Beth-baal-meon	Josh 13.17							
= Beth-meon	Jer 48.23							
= Beon	Num 32.3, Jub.29.10							
Baal-peor	Num.25.3; Hos 9.10; Onom.44.12			cf Beth-peor				
Bamoth-baal	Num 21.10, 22.41, Josh 13.17, cf Mesha 27 Beth Bamoth			Kh el-Queiqiyé (Musil)				Kh.Libb (Dearman)
Bela (= Zoar)	Gen 14.2,8, Onom 42.1							
Beer(-elim)	Num.21.16; Is.15.8				in w. et-temed			
Beth-arbel (Arbela)	Hos 10.14, Onom 14.18		Irbid	Irbid		Irbid		

Biblical (RSV), classical name	*References*	*Robinson & Smith 1841*	*G. A. Smith 1915, 1936**	*F.-M. Abel 1938*	*Simons 1959*	*A. Aharoni 1979*	*B.MacDonald 2000*	*Others*
Beth-diblathaim (= Almon-diblathaim)	Jer 48.22, Mesha 20			Deleilat esh-Sherqīyeh				Kh.et-Teim? Um al Walid? Nitl? ((Dearman)
Beth-gamul	Jer 48.23		Al Jema'il	Kh.el-Ǧumeil		Kh. el-Jumeil	Kh.al-Jumeil	Jemeil (Glueck)
Beth-haram/ haran (Livias)	Num 32.36. Onom 48.13. Ptol.V.15.6		T. Rāmeh	Er-Rāmé	t.er rāmeh/ t. iktanū	T. Iktanû	T. ar-Rama	T.Iktanu (Glueck)
Beth-jeshimoth	33.49, Josh 12.3, 13.20, Ez. 25.9, Onom.48.6		Suweimeh*	Sweimé	t.'adeimeh?		T. al-'Azeimeh	T.'Azeimeh (Glueck)
(Beth-)nimrah	Num 32.3,32,36, Josh 13.27, Onom 44.16		T.Nimrin*	T. Nirmrin	t. bleibil	T. el-Bleibil	T. Nimrin	
Beth-peor	Dt 3.26,29, 4.46, 34.6, Josh 13.20, Onom. 48.3		Esh-Sheik Jayel*	Kh es-Šeih Ǧāyel		Kh. Fāghûr		Kh. el-Mehatta (Henke)
Beth-saida (Julias)	Mk 6.45 etc., Pliny NH 5.21; Onom 58.11; Ptol.V.15.3	Et-Tâbighah	Et-Tell	Et-Tell, Kh el-'Arağ				
Betonim	Josh 13.26, Onom.48.11			Kh. Batné	h.batneh	Kh. Batneh	Kh.Batneh	
Bezer (= Bosor?)	Bezer Dt 4.43, Josh 20.8, 21.36, 1 Chr 6.63, 78; Mesha 27			Prob. Umm el-'Amad		Umm el-'Amad	T. Jalul *or* Umm al-'Amad	Jalul (or Um al 'Amad) (Dearman)
Bosor (= Bostra)	1 Macc.5.26, Onom.12.14 etc		Busr al-Harīri	Busr el-Hariri				
Bostra (Syria)	1 Macc.5.26.Tab. Peut.; Not.Dig. Or.37.1,21; Ptol. V.16.4; Onom. 12.14, 46.8–10, 84.9		Bosra	Bosra eski Šam				
Bozrah (Edom)	Gen.36.33, Am.1.12,Jer.49.13,22, Is.34.6, 1 Chr.1.44, Onom.46.11	El-Busaireh	El-Buseira*	Buseira	el-buseira	Buseirah	Busayra	
Bozrah (Moab) (= Bezer?)	Jer.48.24							
Callirrhoe	Jos.BJ.1.657; Ant.17.171; Plin. NH.V.16; Ptol.V.15.6							Ain ez-āra (Brünnow)

Biblical (RSV), classical name	*References*	*Robinson & Smith 1841*	*G. A. Smith 1915, 1936**	*F.-M. Abel 1938*	*Simons 1959*	*A. Aharoni 1979*	*B.MacDonald 2000*	*Others*
Canata (Kenath)	Jos. BJ.1.367; Ant.15.112; Ptol. V.14.18; Onom.112.20		el-Kerak, or Kunawāt	Kérak-Kanata, or Qanawāt				
Carnaim	1 Macc.5.26,43; 2 Macc.12.21,26; Onom.112.3			Seih Sa'ad	sheih sa'ad	Sheikh Sa'd		
Capitolias	Tab.Peut.; Ptol. V.1.18		Beit Ras	Beit Rās				Beit Ras (Kruse)
Charax (or, palisaded camp)	2 Macc.12.17			El-Kerak (Gilead)				Araq el-Emir (Goldstein)
Chaspho (= Caspin?)	1 Macc.5.23,36		El-Mezeirib	Kusfin *or* el-Muzeirib	el-mzeirīb			
Danaba	Onom.76.11							
Dannea	Onom.76.10			Kh ed-Denn				
Dathema	1 Macc.5.9			T.Hamad nr Šeih Miskīn	t.hamad		Ar-Ramthat?	
Dibon/Dibon-Gad	Num.21.20, 33.45, 32,3,4; Josh 13.9,17; Is.15.1, Jer.48,18,22. Mesha 21,28. Onom 76.17		Dhībān	Dibān	dībān	Dhibân	Dhiban	Dhiban (Dearman)
Dimon (cf Madmen)	Is.15.1		Medeiyineh		h.dimneh		Kh. ad-Dimnah	Ed-Dimnah (Worschech, Knauf)
Dinhabah (=Dannea?)	Gen.36.32, 1 Chr.1.43; Onom.96.9			El-Basta?				
Dion	Pliny NH.V.18; Ptol. V.14.18							
Di-zahab (= Mezahab, Gen.36.39; Meyer)	Dt.1.1, Onom.114.1		Dahab	Ed-Dheibé (Musil)		Dhahab?		Mina ed-Dahab (on G. of Aqabah) (Burckhardt)
Edrei, Rom. Adraene	Num,21.33, Dt.1.4, 3.1,10, Josh.12.4, 13.12,31. Onom.12.13, 84.8		Ed-Dera'ah	Der'ā	der'ā	Der'ā	Dera'a	
Eglath-shelishiyah, Eglaim	Is.15.5,8; = Agalla, Jos.Ant.13.15.4. Onom. 36.20			Rugm el-Ğilimé Musil, Kh.Ğalğul	r. el-ğilime			El-Bleidah (Worschech, Knauf)
Elath	Dt.2.8; 1 K.9.26; 2 Chr.8.17; 2 K.14.22; 16.6; 2 Chr.26.2; Onom.34.25	Ailah, 'Akabah	'Akabah	1 km WNW of Aqaba		'Aqabah	Tall al-Khalayfi	

Biblical (RSV), classical name	*References*	*Robinson & Smith 1841*	*G. A. Smith 1915, 1936**	*F.-M. Abel 1938*	*Simons 1959*	*A. Aharoni 1979*	*B.MacDonald 2000*	*Others*
Elealeh	Num.32.3,37; Is.15.4; 16.9; Jer.48.34; Onom.84.10		El-ʿAl	El-ʿAl	el-ʿal	El-ʿĂl	El-ʿAl	
Ezion-geber	Num.33.35; Dt.2.8; 1 K. 9.26; 22.48; 2 Chr.8.7; 20.36; Onom.36.1, 62.13	El-Ghudyan	Ain el-Ghudyan	On NW coast of Gulf of Aqabah, nr Um Ras Ras (Frank)	t.keleifē	T.el-Kheleifeh	Jezirat Faraʿun?	T. el-Kheleifeh (Glueck); Jezirat Far'on (Rothenberg; Flinder)
Gaia	Onom.62.18			El-Ği				El-dschi (Brünnow)
Gadara	Mt.8.28; Pliny NH. V.18.34; It.Ant.197.6, 198.10; Tab.Peut.; Ptol.V.14.18; Onom.74.10		Mukēs, = Umm Qeis*	Umm Keis				Umm Qeis (Seetzen)
Gadda	Tab.Peut.; Not.Dig. Or.37.9,20		Khan nr Qal'at ez-Zerqa	Kh. es-Samra				Chau nr kalʿat ez-zerkā (Thomsen); Kh. Khaw (Rees); el-Hadīd (Desreumaux & Humbert)
Gerasa	Mk.5.1; Lk.8.26; Pliny NH.V.18.74;. Ptol.V.14.18; 16.4; Onom.64.2, 94.27, 102.21		Jerāsh	Ğéraš	ğerash			
Gergesa	Mt.8.28. Onom.74.14		Kersa	Chorsia				El-Kursi (Tzaferis)
Golan in Bashan	Dt.4.43; Josh.20.8; 21.27;; Onom.64.7		Sahem ej-Jaulān			Sahm el-Jōlân?		
Gudgodah	Dt.10.7, Num.33.23; Onom.64.5			W. Gadagad				
Haila (Aela, Aila. Aelana)	Strabo 16.2.30; Pliny NH.V.11; Tab.Peut.; Not. Dig,Or.34.16,30,48, Ptol.V.16.1., Onom 6.17	Mounds at north-east corner of Gulf of ʿAqabah	Mounds WNW of ʿAqabah	El-ʿAqaba		ʿAqabah		T. el-Kheleifeh (Glueck); Aqaba (Whitcomb)

Biblical (RSV), classical name	*References*	*Robinson & Smith 1841*	*G. A. Smith 1915, 1936**	*F.-M. Abel 1938*	*Simons 1959*	*A. Aharoni 1979*	*B.MacDonald 2000*	*Others*
Hatita	Not.Dig,Or.37.30,31		Kh. es-Samra*	El-Hadīd-Qala'at Zerqa				Al-Hadad, Kal'at ez-zerkā (Brünnow); Kh. as-Samra (Parker, Desreumaux & Humbert)
Havara, Avara	Ptol.V.16.4; Not.Dig. Or.34.12,25 Tab.Peut.			Kh. el-Khaldé				Humeimeh (Musil) El-kwēra (Thomsen)
Helam (= ? Alema)	2 Sam.10.16, 1 Macc.5.26		'Almā	'Alma in Hauran	= alema			
Heshbon, Esbu, Esbuta	Num.21.25; 32.3,37; Josh 21.39, 13. 17,26. Cant.7.4; Is.15.4; 16.8; Jer.48.2,34–45; 49.3. Ptol.V.16.4; Onom.84.4		Hesbān	Hesbān	hesbān	Hesbân	Tall Hisban	
Holon	Jer.48.21			'Aleyan? Kfar Abu Khînân?				
Horonaim	Is.15.5; Jer.48.3,5; Mesha 32			El-'Araq (Musil)		El-'Irâq?	Al-Karak	Ai, e. of Kathrabba, *or* T.Meidan? (Dearman); ed-Deir (Worschech, Knauf)
'Ir Moab /city in the middle of the valley	Num.22.36; Dt.2.36; Josh.13.9,16; 2 S.24.5			El-Medeiyiné	h. el-medeiyineh			
Iye-abarim	Num.21.11–12, 33.44			Mahaiy		El-Medeiyineh?		
Jabbok river	Gen.32.23; Onom.102.19		W. Zerkā	Nahr ez-Zerqā	nahr ez-zerqā		w. az-Zarqa	
Jabesh-gilead	Jud.21.8,14; 1 S. 11.1–10; 31.11–12; 2 S.2.4; 21.12; 1 Chr.10.11. Onom.110.11			Deir Halāwa	t. maqlub	T. el-Maqlûb	Tall al-Meqbereh Tall al-Maqlub, or Abu al-Kharaz	T.Meqbereh, T. abu Kharaz (Glueck)

Biblical (RSV), classical name	*References*	*Robinson & Smith 1841*	*G. A. Smith 1915, 1936**	*F.-M. Abel 1938*	*Simons 1959*	*A. Aharoni 1979*	*B.MacDonald 2000*	*Others*
Jahaz, Jahzah	Num.21.23; Dt.2.26,32; Jud.11.20; Josh.13.8; 1 Chr.6.63; Is.15.4; Jer.48.21. Mesha 20. Onom.104.11		Umm el-Walīd	Ğalūl? Umm el-Walid (Musil).	h. libb?	Kh. el-Medeiyineh?	Kh al-Mudayna in W. ath-Thamad?	Medeiyineh on the W. Themed (Dearman)
Jazer	Num.21.24,32; 32.1,3,35; Josh.17.25; 21.39; 1 Chr.6.65; 2 Sam.24.5; 1 Chr.26.31. Onom.104.13		Beit Zerah	Kh.Ğazzir	h. ğazzir	Kh. es-Sâr?	Kh. Jazzir	Kh.Sar (Seetzen, Merrill); Yajuz (Oliphant); Kh. es-Sireh (Landes)
Jogbehah (Sela)	Num.32.35; Jud.8.11		Al-Jubeihāt, el-Kebireh*	Ğubeihat	el-ağbeihat	El-Jubeihât	Al-Jubeihah	
Joktheel	2 K.14.7	Sela, Petra				(Sela)	Es-Sela‘	
Jotbathah	Dt.10.7; Num.33.33			Et-Tāba		Tâbeh?		
Kamon	Jud.10.5. Polyb.5.70; Onom.116.20		Kumeim/ Kamm	Qamm in ‘Ajlun				
Karakmoba	Ptol.V.16.4; Madeba mosaic		El-Kerak*					el-kerak (Thomsen)
Kedemoth	Dt.2.26; Josh.13.18; 21.37; 1 Chr.6.64; Onom.114.5–6		El-Meshreik	Qasr ez-Za‘feran, or Kh. er-Remeil	qasr ez-za- ‘ferān	‘Aleiyân?	Qasr Saliya (or near)	Saliya (Van Zyl, Dearman)
Kenath (= Nobah), Kanatha	Num.32.42; 1 Chr.2.23; Pliny NH.V.18; Ptol.V.14.18; Onom.112.20		Kunawāt	Kérak-Kanata, or rather Qanawāt	kerak-canatha	Qanawât		el-kanawāt (Thomsen)
Kerioth	Jer.48.24. Amos 2.2; Onom.120.1; Mesha 13			Saliyé?		El-Qereiyât	Diban, capital city of Moabites	Kh. ‘Aleiyan (Dearman)
Kir Hareseth, Kir Heres	2 K. 3.25; Is.16.3,11. Jer.48.31,36; cf. Is.15.1; Jer.48.37		El-Kerak		el-kerak	El-Kerak	Capital city of Moabites	Kerak (Glueck)
Kir Moab; Karaka	Is.15.1; 2 Macc.12.17	Kerak	El-Kerak*		el-kerak			= Rabbathmoba = er-Rabbah (Knauf)
Kiriathaim (cf. Shaveh-kiriathaim Gen.14.5)	Num.32.37; Josh.13.19; Jer.48.1,23; Ez.25.9; Mesha 10; Onom.112.14		Kureiyāt	Kh. el-Qureiyat	h. el-qureijāt	Qaryat el-Mekhaiyet?		Al Qureiye (or Jalul, Kh. et Teim, Kh. Muhaiyat) (Dearman)

Biblical (RSV), classical name	*References*	*Robinson & Smith 1841*	*G. A. Smith 1915, 1936**	*F.-M. Abel 1938*	*Simons 1959*	*A. Aharoni 1979*	*B.MacDonald 2000*	*Others*
Kiriath-huzoth	Num.22.39			El-Qeryé, NE of Diban				
Laban	Dt.1.1			Kh.Libban				
Lasa	Gen.10.19; Onom.120.7							Callirhoe (Jerome)
Libona	Not.Dig.Or.37.27							Kh.Libb (Thomsen)
Lodebar	Josh.13.26 (Debir); 2 S.9.4–5, 17.27; Am.6.13; Onom.122.26		Ibdar*	Umm ed-Dabar		Umm ed-Dabar?		T.Mghanne (Noth); Kh.Hamid (Mittmann); T. Dober (Edelman)
Luhith	Is.15.5; Jer.48.5; Onom.122.28		Tal'at Heiseh	Kh.Fas, or Kh. Medinet er-Ras			Kathrabba area	Kathrabba (Mittmann)
Machaerus	Jos.BJ.9.171–89; Strabo 16.2.40; Pliny NH 5.16		Mukaur = Mukāwer*	Mkawer				El Mekâwer (Seetzen, Glueck)
Madmen	Jer.48.2		Medeiyineh	Kh.Dimné		Kh. Dimneh?	Kh. ad-Dimnah	
Mahanaim	Gen.32.3; Josh.13.26,30; 21.38; 2 S.2.8,12,28; 17.24,27; 19.33; 1 K.2.8; 4.14;1 Chr.6.65			Kh.Mahé? Tulul ed-Dahab (Dalman)	t. heğāğ	T. edh-Dhahab el-Gharbi	T.adh-Dhahab al-Gharbiyya & ash-Sharqiyya	T. Heggag (de Vaux)
Maged	1 Macc.5.26			T. Miqdad				
Masrekah	Gen.36.36. 1 Chr.1.47. Onom.124.18			J. el-Mušrāq	ğebel mushrāq		Jabal al-Mushraq area	? Meshrāqā s. of Kerak (Knauf)
Mattanah	Num.21.18. Onom.126.14			El-Medeiyiné at head of W. Wali	el-medeiyineh			Kh. el-Medeiyineh in W. eth-Themed (Glueck)
Medeba	Mesha 8; Num.21.30; Josh.13.9,16; 1 Chr.19.7; Is.15.2; 1 Macc.9.31; Ptol. V.16.4; Onom.128.19		Mādebā	Madaba	mādebā	Mâdeba	Madaba	Medeba (Dearman)
Mephaath	Josh.13.18, 21.37. 1 Chr.6.79. Not.Dig. Or.37.8,19; Onom.128.21–2			Kh.Nefa'a	h. nef'ah	T. Jâwah	Umm ar-Rasas	Umm er-Rasas (Piccirillo)

Biblical (RSV), classical name	*References*	*Robinson & Smith 1841*	*G. A. Smith 1915, 1936**	*F.-M. Abel 1938*	*Simons 1959*	*A. Aharoni 1979*	*B.MacDonald 2000*	*Others*
Mibsar, Mabsara (=Bozrah)	Gen.36.42. Ps.108.11. Onom.124.20			Buseira	el-buseira		Busayra	
Minnith	Jud.11.33; Ez.27.17; Onom.132.1						In Amman area	Umm el-Basatin (Umm el-Hanafish)
Misor ('tableland')	Dt.3.9, 4.43.Josh. 13.9,16; Jer.48.21; Onom.128.17		'Mishor of Medeba', = s. half of Belqa'.	From Medeba (?) in n. to Arnon in s.	Plain e. of Jordan, s. of Gilead	Plain from Medeba to Dibon	Plateau between Arnon and Heshbon	
Mizpah	Gen.31.49		Sūf	Kh. Ĝel'ad?				
Mizpeh of Moab	1 S.22.3			Rujm el-Mešrefé				
Moa	Medeba mosaic map			El-Mahallé				Moya 'Awād (Donner)
Moahile	Not.Dig.Or.34.29							El-kerak? (Thomsen)
Nadabath	1 Macc.9.37			Kh.et-Teim?	Gk version of Nebo			
Nahaliel	Num.21.19		W. Zerqa Ma'in*	W. el-Habīs (upper W.Zerqa Mā'īn)				
Nebo	Num.32.3,38; Is.15.2; Jer.48.1,22. Mesha 14–18. Onom.136.12		Beit Nūba Jebel Nebā	Kh. el-Mhayiet	h.el-muhājit	Kh.'Ayûn Mûsā?	Kh. al-Mukhayyat	Kh. el Muhaiyat (or Kh 'Ayun Musa,
Negla, Necla	Tab.Peut.; Ptol. V.16.4							'Ain Nejl; 'ain nedschdel (Thomsen)
Nimrah	Num.32.3						Tall Nimrin	
Nimrim, waters of	Is.15.6; Jer.48.34. Onom.138.21		W. Nimeirah	Rujm en-Numeira	seil en-numeirah		W. an-Numayra	Seil & Ghor en-Numeirah (Worschech, Knauf)
Nobah (v. Kenath)	Num.32.42					(v. Kenath)		
Oboth	Num.21.10; 33.44			El-Weibé	el-weibeh		Between W. Khanazir & Bab adh-Dhra'	
Pella	Jos.BJ.1.104; Pliny NH.V.18; Ptol. V.14.18; Onom.14.19, 32.6, 110.13		Kh. Fāhil, Tabagat Fahil*	Fahil		Kh. Fahil		
Penuel	Gen.32.23,31; Jud.8.8,13,17; Onom. 168.4, 170.2		Jebel Osha' = Jebel Yusha*	Tulul ed-Dahab	tulul ed-dahab	T. edh-Dhahab esh-Sherkîyeh	Tall Dayr 'Alla	

Biblical (RSV), classical name	*References*	*Robinson & Smith 1841*	*G. A. Smith 1915, 1936**	*F.-M. Abel 1938*	*Simons 1959*	*A. Aharoni 1979*	*B.MacDonald 2000*	*Others*
Petra (Sela, LXX; Arke, Rekeme, Jos. Ant.4.82,161)	Strabo XVI.4.24; Ptol.V.16.4; Tab. Peut.; Onom.36.13, 142.7, 144.7	W.Mûsa	W. Mūsa	Petra=Sela' =Joktheel =Umm el-Biyâra				
Philadelphia	Strabo XVI.2.34; Pliny NH.V.18; Ptol. V.14.18; Onom.16.15, 24.2, 102.21, 146.7		'Ammān	'Ammān				'ammān (Thomsen)
Praesidium	Tab.Peut.;Not.Dig. Or.34.35,41; Madeba mosaic map			El-Kitara				El-mesadd in W. Itm (Thomsen); Qasr el-Feifeh (Frank); Qasr el-Kitharā (Glueck); Kh. el-Khalde (Graf, Parker)
Punon	Gen.36.41. Num.33.42,43; Onom.168.8			Feinān	feinān	Feinân	Faynan	Feinan (Glueck)
Qrhh	Mesha 3, 21–25							Quarter (acropolis?) of Dhiban (Dearman)
Rabbah of the Ammonites	Dt.3.11; 2 Sam.11.1; 12.26;Jer.49.3; Ez.21.25		Ammān		'amman	Ammân	Amman Citadel, Jabal al-Qal'ah	Jebel Qal'ah, Amman
Rabbath-moba (Areopolis)	Ptol.V.16.4; Onom.16.17, 36.24, 124.16–17; Not.Dig. Or.37.5; Tab.Peut. Rababatora	Rabba = Ar of Moab, Areopolis	Rabba	Er-Rabba				er-rabba (Thomsen)
Ramath-mizpeh	Josh.13.26			prob. Kh.Gel'ad	h.sār?	Kh. Jel'ad?		
Ramoth-gilead	Dt.4.43; Josh.20.8; 21,38;1 K.4.13; 22.3–26; 2 Chr.18; 2 K. 8.28; 9.1–14;1 Chr.6.65; 2 Chr.22.5; Onom.144.4–5		Remtheh	Er-Ramta (Jos. Aramatha) Hosn	t.er-rāmīt	T. Rāmîth	Tall al-Husn	T. Husn (Dalman, Albright); T. Ramith (Glueck)
Raphon	1 Macc.5.37; Pliny NH.V.18 Raphana			Er-Refi	er-rāfeh	Er-Râfeh		

Biblical (RSV), classical name	*References*	*Robinson & Smith 1841*	*G. A. Smith 1915, 1936**	*F.-M. Abel 1938*	*Simons 1959*	*A. Aharoni 1979*	*B.MacDonald 2000*	*Others*
Rehoboth	Gen.36.37; 1 Chr.1.48; Onom.142.13–14. Not.Dig, Or.34.11,27, Robatha			Kh.Musrab				Ras er-Rihāb (Zwickel); W. al- Ghuwayr region (Knauf)
Rekem (=Petra, Jos.)	Num.31.8; Onom.144.7; 36.13; 144.7		W. Mūsa			Umm el-Bayyâreh?		
Rogelim	2 Sam,17.27, 19.32; Onom.146.9			Er-Ruğeili				
Salecah	Dt.3.10; Josh.13.11; 12.4,5; 1 Chr.5.11; Onom. 154.15, 156.2		Salkhad	Salhad		Salkhad		
Salmonah	Num.33.41. Not.Dig. Calamona			Bir Madkur				
Sartabe (Alexandrion)	Jos.Ant.14.49; BJ 1.167		Kurn Surtubeh	Qarn Sartabé				
Sebam, Sibmah	Num.32.3, 38; Is.16.8–9; Jer.48.32		Sūmia	Qurn el-Kibš	qarn el-qibsh			
Sela	2 K.14.7; 2 Chr.25.12; Isa.42.11; Jer.49.16; Obad.3	= Petra, W. Musa	W. Musa	Umm el-Biyāra	umm el-biyārah	Es-Sela'	As-Sila'	Umm el-Biyarah (Glueck)
Succoth	Gen.33.17; Josh.13.27; Jud.8.5,16; 1 K.7.46; Ps.60.8		T. Deir 'Alla	T. Ahsās	t. deir 'allah	T.Deir 'Allā	Tall al-Khisas	T. Ahsas (Franken); T. Umm Hamad (Glueck)' T. es-Sa'idiyeh (Ottoson)
Teman	Gen.36.34; Jer.49.7,8,20; Ez.25.12–14; Am. 1.12; Hab.3.3; Obad.9; Onom.96.20			(rejects Glueck's identification with Tawilan)		Tawilan	Entire land of Edom	n. part of Edom around Bozrah (Knauf)
Thantia, Thainatha	Tab.Peut; Not.Dig. Or.37.29			Umm el-Ğimal				el-chab (Thomsen); Thughrat al-jubb (Kennedy)
Thisbe	1 K. 17.1; Onom.102.6			El-Istib/Lisdib				
Thoana	Ptol.V.16.4; Peut. Thornia;	Dhâna	Dat Ras (Thorma)	Et-Tuané				tawāne (Brünnow)

Biblical (RSV), classical name	*References*	*Robinson & Smith 1841*	*G. A. Smith 1915, 1936**	*F.-M. Abel 1938*	*Simons 1959*	*A. Aharoni 1979*	*B.MacDonald 2000*	*Others*
Toloana, Toloha	Not.Dig.Or.34.34			Et-Telāh				tlah (Musil)
Tophel	Dt.1.1	Tūfîleh		Et-Tafilé	et-tafileh	Et-Tafîleh?		
Tyros (= wadi sir)	Jos.Ant.12.233							ʻarak el-emīr (Thomsen)
Valtha	Not.Dig.Or.37.33							umm el-waīd (Thomsen)
Zaphon	Josh.13.27		=Amatha, T.Hammi*	T. es-Sa'idiyé?	t. el-qōs	T. es-Sa'īdîyeh	Tall as-Sa'idiyya	T. el-Qôs (Glueck)
Zarethan	1 K.4.12; 7.46; Onom. 160.18			Qarn Sartabé?	near qarn sartabeh	T. Umm Hamâd?		
Zered, brook	Dt.2,13,14,18; Num.21.12; Onom.92.10	W. el-Ahsy	W. ʻAin Feranjy, or Seil Sʻaideh; W. es-Sultani (upper W.Mojib)*					
Zereth-shachar	Josh.13.19			Zārāt	ʻon the hill of the Jordan valley'	Ez-Zārât	Boz al-Mushelle?	el-Zarat (Boling & Wright)
Zoar, Zoara	Gen.13.10; 14.2,8; 19.20; Dt.34.3; Is.15.5; Jer.48.34; Ptol.V.16.4; Not.Dig. Or.34.7, 26;Onom. 42.1; Madeba mosaic map;	At mouth of w.Kerak	El-Keryeh			Es-Sâfi	At SE plain of Dead Sea	Tawāhīn as-Sukkar (Glueck, Rast & Schaub, Knauf)
Zodocatha, Zadagatta	Tab.Peut,; Ptol. V.16.4 (Zaanatha); Not.Dig.Or.34.8, 24;.	Ŭsdakah		Es-Sadaqa				Sadeqah (Glueck)
Ziza	Ptol.V.16.4; Not.Dig. Or.37.7,16							El kastal n. of zīzā (Brünnow, Thomsen)

BIBLIOGRAPHY

Abel, F.-M., 1933, 1938, *Géographie de la Palestine* I–II (Paris: Gabalda)

Ackerman, S. (ed.), 1998, *Humphrey Cole: Mint, Measurement and Maps in Elizabethan England* (British Museum Occasional Paper 126)

Addison, W., 1951, *Worthy Dr Fuller* (London: Dent)

Adrich (Adrichomius), Christian, 1593, *Theatrum terrae sanctae et biblicarum historiarum cum tabulis geographicis aere expressis* (Koln: Birckmann)

Aharoni, Y., 1966, *The Land of the Bible* (London: Burns and Oates)

Akerman, J. R., 1995, 'From books with maps to books as maps: The editor in the creation of the atlas idea', in J. Winearls (ed.), *Editing Early and Historical Atlases* (Toronto: University of Toronto Press), 3–48

Allatius, Leo, 1653, *ΣΥΜΜΙΚΤΑ, sive opusculorum Graecorum et Latinorum vetustiorum ac recentium libri duo*, ed. B. Nihusius (Coloniae Agrippinae, apud I. Kalcovium)

Allgemeine deutsche Biographie, 1875–1912 (Leipzig: Duncker und Humblot)

Almagià, R., 1927, 'La carta della Palestina di Gerard Mercatore 1537', in *Fondo Italiano per lo Studio della Palestina* I (Firenze: Istituto Geografico Militare)

Almagià, R., 1951, 'On the cartographic work of Francesco Rosselli', *IM* 8, 27–34

Almagià, R., 1951, 'A hitherto unknown map of Palestine', *IM* 8, 34 and plate

Alt, A., 1936, obituary of Hermann Guthe, *ZDPV* 59, 177–80

Alt, A., 1942, obituary of Hans Fischer, *ZDPV* 65, 5–8

D'Anville, Jean-Baptiste Bourguignon, 1732, 'Patriarchatus Hierosolymitanus' (engraving)

D'Anville, Jean-Baptiste Bourguignon, 1767, 'La Palestine' (engraving)

Arias Montanus, Benedictus, 1572, *Phaleg sive de gentium sedibus primis, orbisque terrae situ liber* (Antverpiae: Plantin)

Avi-Yonah, M., 1925, 'Map of Roman Palestine 1:250,000', *QDAP* 5, 139–93

Avi-Yonah, M., 1954, *The Madeba Mosaic Map with introduction and commentary* (Jerusalem: Israel Exploration Society)

Avi-Yonah, M., 1976, *Gazetteer of Roman Palestine, Qedem* 5 (Jerusalem: The Institute of Archaeology, The Hebrew University of Jerusalem; CARTA)

Bagrow, L., 1950, 'Essay of a catalogue of maps and incunabula', *IM* 7, 106–09

Bagrow, L., 1964, *History of Cartography* (revised and enlarged by R. A. Skelton) (London: Watts)

Bailey, J. E., 1874, *The Life of Thomas Fuller, D. D., with Notices from his Books, his Kinsmen and his Friends* (London: Pickering)

Baker, J. N. L., 1937, *A History of Geographical Discovery and Exploration*, 2nd ed. (London: Harrap)

Barber, P., 2006, 'Medieval maps of the world', in P. D. A. Harvey (ed.), *The Hereford World Map: Medieval world Maps in their Context* (London: British Library), 1–44

Barker, P., 1998, 'Humphrey Cole's Map of Palestine', in S. Ackerman (ed.), *Humphrey Cole: Mint, Measurement and Maps in Elizabethan England* (London: British Museum Occasional Paper 126), 97–100

Bartlett, J. R., 1989, *Edom and the Edomites* (JSOTsup 77: Sheffield: Sheffield Academic Press in association with the Palestine Exploration Fund)

Beazley, C. R., 1926, 'A new find of Mercator maps', *GJ* 68.6, 527–28

Becker, F., 1913, 'Zur Kartographie Palästinas', *ZDPV* 36, 198–202

Ben-Arieh, Y., 1979, *The Rediscovery of the Holy Land in the Nineteenth Century* (Jerusalem: The Magnes Press, The Hebrew University, Israel Exploration Society; Detroit: Wayne State University Press)

Benedictus Arias Montanus, 1572, *Phaleg, sive de gentium sedibus primis, orbisque terrae situ, liber* (Antverpiae: C. Plantinus)

Bernard, J. H. (trans)., 1894, *Account of the Holy Land written about* A.D. *1350 from MS D.4.7 in Trinity College Dublin Library* (London: PPTS vol. 6)

Bible, Geneva Version, 1582 (London: Barker)

Bible, The Holie (Bishops' Bible), 1572 (London: R. Jugge)

Bible, New Testament (English), 1566 (London: R. Jugge)

Bituri, P. L., 1651, *Notitia Dignitatum Imperii Romani* (Paris)

Blaeu, Gulielmus, 1631, *Appendix Theatri A. Ortelii et Atlantis G. Mercatoris* (Amsteldami: apud Guiljemum Bleauv)

Blaeu, J., 1662, *Asia quae est Geographiae Blavianae pars quarta, libri duo, volumen decimum* (Amstelodami: J. Blaeu)

Blaeu, Joan, 2005, *Atlas Maior of 1665* (new ed. introduced by P. van der Kogt) (Köln, London, Los Angeles, Madrid, Paris, Tokyo: Taschen)

Böcking, E., 1839–53, *Notitia Dignitatum et administrationum omnium tam civilium quam militarium, in partibus Orientis et Occidentis* (Bonnae: A. Marcus)

Boeren, P. C., 1980, *Rorgo Fretellus de Nazareth et sa description de la Terre Sainte: Histoire et edition du texte* (Amsterdam, Oxford, New York: North Holland Publishing Co.)

Bonfrerius, J., 1631, 'Onomasticon urbium et locorum sacrae scripturae', appended to *Iosue, Iudices et Ruth commentario illustrati* (Paris: Cremoisy)

Bonfrerius, J., 1657, 'Annotationes in Christiani Adrichomii descriptionem Terrae Sanctae', in B. Walton, *Biblia Sacra Polyglotta* 1, 49–50 (London: T. Roycroft)

Bongarsius, J. (ed.), 1611, *Orientalis Historiae Tomus II* (Hanoviae: apud haeredes I. Aubrii)

Boorsch, S., 2004, 'The case for Franceso Rosselli as the engraver of Berlinghieri's *Geographia*', *Imago Mundi* 56.2, 152–60

Bowersock, G. W., 1983, *Roman Arabia* (Cambridge, Mass., and London: Harvard University Press)

Bowersock, G. W., 2006, *Mosaics as History: The Near East from late Antiquity to Islam* (Cambridge, MA, and London, England: Belknap Press of Harvard U.P.)

von Breitenbach, Bernhard, 1486, *Sanctarum Peregrinationum in Montem Syon ad venerandum Christi sepulchrum in Jerusalem atque in montem Synai ad divam virginem et matirem Katherinam opusculum* (Mainz: E. Reuwich)

Breydenbach, 1536, *Peregrinatio ad terram sanctam* (Vitembergiae: Nicola Schir)

Brocardus, 1537, 'Locorum Terrae Sanctae exactissima descriptio', in S. Grynaeus, *Novus orbis regionum ac insularum veteribus incognitarum una cum tabula cosmographica* (Basileae: apud Io. Hervagium), 298–329

Brocardus, monachus, ed. J. Bonfrerius, 1711, *Descriptio terrae sanctae*, recensuit J. Clericus (Amsterdam: apud R. and G. Wetstenios)

Brooker, C. H., and Knauf, E. A., 1988, 'Notes on Crusader Transjordan (Pls 14–15)', in review of J. Prawer, *Crusader Institutions* (Oxford: Clarendon, 1980), *ZDPV* 104, 184–88

Le Brun [Bruyn], Corneille (C. van Bruyn), 1700, *Voyages au Levant et dans les principales parties de l'Asia Mineure* (Delft); Eng. trans., 1702, *A Voyage to the Levant, or, Travels in the principal Parts of Asia Minor* (London); original version, 1698, *Reizen van C de Bruyn* (Delft)

Brünnow, F. F. E., obituary notice of, 1892, *Monthly Notices of the Royal Astronomical Society* 52, 230–33

Brünnow, R. E., and von Domaszewski, A., 1904–1909, *Die Provincia Arabia auf Grund zweier in den Jahren 1897 und 1898 ubernommenen Reisen und der Berichte früheren Reisender, vols I–III* (Strassburg: K. J. Trübner)

Burchard (Brocardus) of Mount Sion (c. 1280–83), *Descriptio Terrae Sanctae*, v. sub Canisius, H., Laurent, J. C. M., and Stewart, A

Buckingham, J. S., 1821, *Travels in Palestine, through the countries of Bashan and Gilead, east of the River Jordan* (London: Longman, Hurst)

Burckhardt, J. L., 1822, *Travels in Syria and the Holy Land* (London: J. Murray)

Campbell, T. A., 1981, review of G. Grosjean, *The Catalan Atlas of 1375*, *IM* 33 (1981), 115–16

Campbell, T. A., 1987, *The Earliest Printed Maps 1472–1500* (London: The British Library)

Canisius, Henricus, 1604, 'Descriptio Terrae Sanctae auctore Burchardo de Monte Sion, nunc primum in publicum data', in *Antiquae Lectionis Tomus VI* (Ingoldstadii: apud A. Angermarium)

Cellarius, Christophorus, 1698, *Historia antiqua* (Jena: sumptu I. Bielkii)

Cellarius, Christophorus, 1701, 1793, 1706 and later editions, *Notitiae Orbis Antiqui sive Geographiae plenioris Tomus Alter, Asiam et Africam antiquam exponens* (Leipzig: apud Ioh. F. Gleditschii; Cambridge; Amsterdam)

Cellarius, Daniel, 1578, *Speculum orbis terrarum* (Amsterdam; de Iode)

Churchill, Awnsham and John, 1704 (vols 1–4), 1732 (vols 5,6), *Collection of Voyages and Travels, some now first printed from original manuscripts* (London: J. Walthoe)

Cobbing, F., 2005, 'The American Palestine Exploration Society and the Survey of Eastern Palestine', *PEQ* 137.1, 9–21

Committee of the PEF, 1873, *Our Work in Palestine being an Account of the different Expeditions sent out to the Holy Land* (London: R. Bentley)

Conder, C. R., and H. H. Kitchener, 1880, *Map of Western Palestine in 26 sheets from Surveys conducted for the Committee of the [PEF]. . . .1872–77* (London: Committee of the PEF)

Conder, C. R., and H. H. Kitchener, 1881–83, *Memoirs of the Survey of Western Palestine* (3 vols, London: Committee of the PEF)

Conder, C. R., 1889, *The Survey of Eastern Palestine. Memoirs of the Topography, Orography, Hydrography, Archaeology etc. Volume 1: The 'Adwan Country* (London: Committee of the PEF)

Coverdale, Miles, 1535, *Biblia. The Bible: that is, the Holy Scriptures of the Old and New Testaments, faithfully translated into English* (Zurich?: Froschover); facsimile *The Coverdale Bible 1535, with an introduction by S. L. Greenslade*, 1975, Folkestone: Dawson

Cox, Edward G., 1938, 'Maps and atlases — a selected bibliography', in id., *Reference Guide to the Literature of Travel, including Voyages* (Seattle: University of Washington), II, 396–416

Crane, N., 2002, *Mercator, the Man who mapped the Planet* (London: Weidenfeld and Nicolson)

Cruickshank, J. L., 2004, 'Palestine of the Crusades: almost an Ordnance Survey map', *Streetlines* 71, 10–15

Cuntz, O., 1929, *Itineraria Antonini Augusti et Burdigalense (cum tabula geographica)* (Itineraria Romana I, Leipzig: Teubner)

Daniell, D., 2003, *The Bible in English* (New Haven & London: Yale University Press)

Davies, H. W., 1911, *Bernhard von Breitenbach and his Journey to the Holy Land 1483–4: A Bibliography* (London: J. & J. Leighton)

Dearman, A. (ed.), 1989, *Studies in the Mesha Inscription and Moab* (Atlanta, Georgia: Scholars Press)

Delano-Smith, C., 1987, 'Maps in Bibles in the sixteenth century', *The Map Collector* 39, 3–14

Delano-Smith, C., 1990, 'Maps as art and science: maps in sixteenth-century Bibles', *IMi* 42, 65–83

Delano-Smith, C., 1991, 'Geography or Christianity? Maps of the Holy Land before A.D. 1000', *JTS* 42, 142–52

Delano-Smith, C., and E. M. Ingram, 1991, *Maps in Bibles 1500–1600: An Illustrated Catalogue* (Genève: Librairie Droz)

Delano-Smith, C., and E. M. Ingram, 1994, 'La carte de la Palestine', in M. Watelet, ed., *Gerard Mercator cosmographe* (Antwerp: Fonds Mercator), Pt. IV, 268–83

Delisle, G., 1700–1724, *Atlas of the World* (Paris)

Desreumaux, A., and Humbert, J.-B., 1981, 'Hirbet es-Samra', *ADAJ* 25, 33–84

Destombes, M., 1964, *Mappaemondes A.D. 1200–1500* (Amsterdam: N. Israel)

Dilke, O. A. W., 1985, *Greek and Roman Maps* (London: Thames and Hudson)

Dilke, O. A. W., 1987, 'Itineraries and geographical maps in the early and late Roman Empire', and 'Cartography in the Byzantine Empire', in J. B. Harley and D. Woodward (eds), *The History of Cartography, I* (Chicago: University of Chicago Press), 234–75

Donner, H., 1982, 'Mitteilungen zur Topographie des Ostjordanlandes anhand der Mosaikkarte von *Mādebā*: Eine Nachlese', *ZDPV* 98, 174–91

Donner, H., 1984, 'Transjordan and Egypt on the mosaic map of Madeba', *ADAJ* 28, 249–57

Donner, H., 1992, *The Mosaic Map of Madeba: An Introductory Guide* (Kampen: Kok Pharos)

Dürst, A., 1991, 'Zur Wiederauffindung der Heiligland-Karte von ca. 1515 von Lucas Cranach dem Alteren', *Cartographica Helvetica*, Jan. 1991, Heft 3, 22–27

Dürst, A., 1998, 'The Map of Europe', in M. Watelet (ed.), *The Mercator Atlas of Europe. Facsimile of the Maps by Gerardus Mercator contained in the Atlas of Europe, circa 1570–1572* (Pleasant Hill, OR.: Walking Tree Press)

Edson, E., 1994, 'Matthew Paris' "other" map of Palestine', *The Map Collector* 66. 1, 18–22

Edson, E., 1997, *Mapping time and Space: How Medieval Mapmakers viewed their World* (London: British Library)

Edson, E., and Savage-Smith, E., 2004, *Medieval Views of the Cosmos* (Oxford: Bodleian Library)

Finkelstein, I., 1979, 'The Holy Land in the Tabula Peutingeriana: a historical-geographical approach', *PEQ* 111. 27–34

Fischer, H., 1939, 1940, 'Geschichte der Kartographie von Palastina', *ZDPV* 62, 169–89; 63, 1–111

Fischer, H., 1913, 'Referat über die moderne Topographie, Siedlungs- und Verkehrsgeographie Palästinas, besonders für die Jahre 1910–1912', *ZDPV* 36, 136–62; 211–19; Taf. 23

Fischer, H., 1890, 'Begleitworte zur Neuen Handkarte von Palästina', *ZDPV* 13, 44–59, Taf. 2

Fischer, H., and H. Guthe, 1895, 2nd ed. 1905, *Handkarte von Palästina zur biblischen Geschichte. Nach den Angaben der Bibel bearb.* 1:200,000. Leipzig: Wagner & Debes

Fischer, H., and H. Guthe, 1911, *Bibelatlas* (revised ed. 1926) (Leipzig: Wagner & Debes)

Freeman-Grenville, G. S. P., R. L. Chapman III and J. E. Taylor, 2003, *The Onomasticon by Eusebius of Caesarea* (Jerualem: Carta)

Fuller, T., 1655, *The History of the University of Cambridge from the Conquest to 1634* (Cambridge); id., ed. by M.Prickett & T. Wright, Cambridge 1840)

Fuller, T., 1650, *A Pisgah-sight of Palestine and the Confines Thereof, with the History of the Old and New Testaments acted thereon* (London: John Williams)

Gavish, D., 1985, 'Innovative cartography in Palestine 1917–1918', *Cartographic Journal* 22, 38–44

Gavish, D., 1994, 'French cartography of the Holy Land in the nineteenth century', *PEQ* 126, 24–31

Gavish, D., 2005, *The survey of Palestine under the British Mandate* (London: Palestine Exploration Fund, RoutledgeCurzon)

Gazetteer of Jordan: Names approved by the United States Board on Geographic Names (Washington D.C.: Defense Mapping Agency, 2nd ed., 2000)

Genesius, J., 1733, *De Rebus Constantinopolitanis a Leone Armenio ad Basilium Macedonem libri quatuor* (Venice: typis J. B. Pasquali)

Goren, H., 1999, 'Heinrich Kieppert in the Holy Land, Spring 1870: Sketches from an exploration-tour of an historical cartographer', in *Antike Welten Neue Regionen: Heinrich Kieppert 1818–1899*, 45–61 (Berlin: Staatsbiblothek)

Goren, H., 2002, 'Sacred but not surveyed: nineteenth-century surveys of Palestine', *IM* 54, 87–109

Goren, H., 2003, *'Zieht hin und erforscht das Land': Die deutsche Palastinaforschung im 19. Jahrhundert* (Schriftenreihe des Instituts fur deutsche Geschichte der Universitat Tel Aviv 23; Göttingen: Wallstein Verlag)

Goss, J., 1990, introd. to *Blaeu's The Grand Atlas of the Seventeenth-Century World* (London: Studio Edition and The Royal Geographical Society)

Graf, D. J. , 1995, 'The Via Nova Traiana in Arabia Petraea', *JRA* 14, 141–67

Graf, D. J., 1999, 'Roman roads east of the Jordan', in M. Piccirillo & E. Alliata (eds), *The Madaba Map Centenary* 1897–1997: Travelling through the Byzantine Ommayad Period (SBF Collatio Maior 40) (Jerusalem: SBF)

Graham, M. P., 1999, 'Guthe, Hermann (1849–1936)', in J. H. Hayes (ed.), *Dictionary of Biblical Interpretation* (Nashville: Abingdon) I, 473–74

Grosjean, G., 1978, *The Catalan Atlas of the Year 1375, edited with commentary* (Dietikon-Zurich: Urs Graf Verlag)

Grynaeus, Simon, 1537, 1555, *Novus orbis regionum ac insularum veteribus incognitarum una cum tabula cosmographica* (Basileae: apud Io. Hervagium)

Gulforde, Richard, 1851, *The Pylgrymage of Sir Richard Guylforde to the Holy Land AD 1556. From a copy from the press of Richard Pynson, ed. by Sir Henry Ellis* (London: Camden Society, 51)

Guthe, H., 1909, 'Die Aufnahme des Ostjordanlandes durch den Deutschen Palastina Vereins', *Mitteilungen d. Ver. f. Erdkunde zu Leipzig*, 35–50

Harley, J. B., and Woodward, D., 1987, *The History of Cartography I: Cartography in Prehistoric, Ancient and Medieval Europe and the Mediterranean* (Chicago and London: University of Chicago Press)

Hart, C. A., 1940, 2nd ed., 1943, *Air Photography applied to Surveying* (London: Longman, Green)

Hart, I. D., 1965, 'The surveys of Palestine', *Royal Engineers Journal* 79.3, 214–26

Hartmann, M., 1890, 'Bemerkungen zu den neuen Haurankarte', *ZDPV* 13, 60–64

Hartmann, R., 1911, 'Die Palastina-Route des itinerarium Burdigalense', *ZDPV* 33, 169–88

Harvey, P. D. A., 1980, *The History of Topographical Maps: Symbols, Pictures and Surveys* (London: Thames and Hudson)

Harvey, P. D. A., 1991, *Medieval Maps* (London: British Library)

Harvey, P. D. A., 1997, 'The Sawley Map and other world maps in twelfth-century England', *IM* 49, 33–42

Harvey, P. D. A., 2001, 'Matthew Paris' maps of Palestine', in Prestwich, M., R. Birtwell and R. Frame (eds), *Thirteenth Century England*, 8 (Woodbridge: Boydell and Brewer)

Harvey, P. D. A., 2002, *Mappa Mundi: The Hereford World Map* (2nd ed., Hereford: Hereford Cathedral)

Harvey, P. D. A., 2003, 'The biblical content of medieval maps of the Holy Land', in Dagmar Unverhan (ed.), *Geschichtsdeutung auf alten Karten: Archäologie und Geschichte* (Wolfenbütteler Forschungen 101, Wiesbaden: Harrassowicz), 56–63

Harvey, P. D. A. (ed.), 2006, *The Hereford World Map: Medieval World Maps and their Context* (London: British Library)

Homann, J. B., 1731, *Atlas Novus Terrarum Orbis Imperia, Regna et Status exactis tabulis geographicis demonstrans* (vol. 1); *Atlas Cosmographicus Major, sistens mundum universum tam coelum quam terram* (vol.2) (Noribergae ex officina Homanniana)

Hornius, G., 1677, *Accuratissima Orbis Antiqui Delineatio sive Geographia Vetus, sacra et profana* (Amsterdam: J. Waesberg)

Hull, E., 1885, *Mount Seir, Sinai, and Western Palestine: being a narrative of a scientific expedition* (London: PEF)

Hull, E., 1886, *Memoir on the Geology and Geography of Arabia Petraea, Palestine, and Adjoining Districts* (London: PEF)

Humbert, J.-B., 1987, 'Khirbet es-Samra, le route et questions de chronologie', *SHAJ* III, 307–10

Huygens, R. B. C., 1986, *Chronique: Guillaume de Tyr* (Corpus Christianum Continuatio Medievalis LXIII, LXIIIA)(Turnhout: Brepols)

Ingram, E. M., 1993, 'The map in the Coverdale Bible: a map by Holbein?', *The Map Collector* 64, 26–31

Ingram, E. M., 1993, 'Maps as readers' aids: maps and plans in the Geneva Bibles', *IM* 45 (1993), 29–44

Jacotin, M., 1810, *Carte topographique de l'Egypte et de plusieurs parties des pays limitrophes levée pendant l'expédition de l'armée française* (Paris: Impr. Royale)

Jaillot, H., 1689, *Atlas Nouveau, contenant toutes les parties du monde* (Paris)

Janssonius, Johannes & Henricus Hondius, 1638, *Atlas Novus* (3 vols) (Amsterdam)

Janssonii Haeredes, 1675, *Atlas Maior* (11 vols) (Amsterdam)

Jansson, J., 1666, *Atlas Contractus, sive Atlantis Majoris Compendium* (Amsterdam: I. Iansonii haeredes)

Johns, J., and Savage-smith, E., 2002, '*The Book of Curiosities*: a newly discovered series of Islamic maps', *IM* 55, 7–24

Jones, Y., 1973, 'British military surveys of Palestine and Syria 1840–1841', *Cartogaphic Journal* 10.1, 29–41

Kallner, D. H., 1944, 'Jacotin's map of Palestine, surveyed during Napoleon's campaign in 1799', *PEQ* 76, 157–63

Karmon, Y., 1960, 'An analysis of Jacotin's Map of Palestine', *IEJ* 10, 155–73, 344–53

Karrow, R. W., 1993, *Mapmakers of the sixteenth century and their Maps* (Chicago: Speculum Orbis Press for the Newberry Library)

Kennedy, D., 1997, 'Roman roads and routes in north east Jordan', *Levant* 29, 71–93

Keuning, J., 1947, 'The history of an atlas, Mercator-Hondius'. *IM* 4, 37–62

Keuning, J., 1948, 'Jodocus Hondius Jr', *IM* 5, 63–71

Keuning, J., 1951, 'The Novus Atlas of Johannes Janssonius', *IM* 8, 71–98

Keuning, J., 1952, 'XVIth century cartography in the Netherlands (mainly in the Northern Provinces)', *IM* 9, 35–63

Kiepert, H., 1841, 'Memoir on the maps accompanying this work', in E. Robinson and Eli Smith, *Biblical Researches in Palestine, Mount Sinai and Arabia Petraea: A Journal of Travels in the Year 1838* (London: J. Murray), vol. III, Appendix B, 29–55

Klostermann, E., 1904, Eusebius: *Das Onomastikon der biblischen Ortsnamen* (Die griechischen christlichen Schriftsteller der ersten drei Jahrhunderte: Eusebius III.1) (Leipzig: J. C. Hinrichs)

Knauf, E. A., 1991, 'Historical studies of toponymy', in J. M. Miller (ed.), *Archaeological Survey of the Kerak Plateau* (ASOR Archaeological Reports 1. Atlanta, GA: Scholars Press), 281–90

Koeman, I. C., 1969, *Atlantes Neerlandici* (Amsterdam: Theatrum Orbis Terrarum); new ed. by P. van der Krogt, 1997, vol.1, *The Folio Atlases published by Gerard Mercator, Jodocus Hondius, Henricus Hondius, Johannes Janssonius and their successors* ('t Goy-Houten, Netherlands: HES Publishers)

Koeman, I. C., 1970, *Joan Blaeu and his Grand Atlas* (Amsterdam: Theatrum Orbis Terrarum)

Krogt, P. van der, 1996, 'Amsterdam atlas production in the 1630s: a bibliographer's nightmare', *IM* 48, 149–73

Kruse, F. C. H., 1854–59, *Ulrich Jasper Seetzen's Reisen durch Syrien, Palästina, Phönicien, die Transjordan-Lander, Arabia Petraea und Unter-Aegypten, heraus gegeben und commentirt von F. Kruse in Verbindung mit... Dr Hinrichs, G. F. H. Müller und mehreren andern Gelehrten, I–IV* (Berlin: G. Reimer)

Kunz, A., 1995, 'Cranach as cartographer: the redisovered map of the Holy Land', *Print Quarterly* 12.2, 123–44

Laborde, L. de, 1830, *Voyage de l'Arabie Petrée par Léon de Laborde et Linant* (Paris: Girard)

Laborde, L. de, 1836, *Journey through Arabia Petraea to Mount Sinai and the excavated city of Petra, the Edom of the prophecies* (London: J. Murray)

Laor, E., and Klein, S., 1986, *Maps of the Holy Land. Cartobibliography of printed maps, 1475–1900* (New York: A Liss; Amsterdam: Meridian)

Laurent, J. C. M. , 1864, 2nd ed., 1873, *Peregrinatores medii aevi quatuor* (Leipzig: J. Hinrichs)

LeStrange, G., 1890, *Palestine under the Moslems...from the works of medieval Arabian geographers* (London: PEF)

Leo Allatius, 1653, *Symmicta* (Col. Agripp. [Köln]: apud Iodoc. Kalcovius)

Lewis, S.,1987, *The Acta of Matthew Paris in the* Chronica Majora (Californian Studies in the History of Art. Berkeley and London: University of California Press, in collaboration with Corpus Christi College, Cambridge)

Lightfoot, J., 1658–78, *Horae Hebraicae et Talmudicae* (6 vols, Cambridge, London, Paris, Leipzig)

Lynch, W. F., 1849, *Narrative of the United States' Expedition to the River Jordan and the Dead Sea* (London: R. Bentley)

MacDonald, B., 2000, *East of the Jordan: Territories and Cities of the Hebrew Scriptures* (ASOR Books, vol.6. Boston, MA: ASOR)

Macpherson, J. R. S. (1892), *Description of Jerusalem and the Holy Land. Fetellus (c. 1130 A.D.)* (London: PPTS vol. 5, 8–54)

Mercator, Gerardus, 1537, map entitled 'Amplissima Terrae Sanctae Descriptio ad Utriusque Testamenti Intelligentiam' (Louvain)

Mercator, Gerardus, 1584, *Cl. Ptolemaei Geographiae Libri octo recogniti iam et diligenter emendati cum tabulis geographicis ad mentem auctoris restitutis ac emendatis* (Coloniae Agrippinae: G. Kempen)

Mercator, Gerardus, 1595, *Atlas sive cosmographicae meditationes de fabrica mundi et fabricati figura* (Duisburg: Clivorum)

Mercator, Gerardus, 1606, *Atlas sive cosmographicae meditationes de fabrica mundi et fabricati figura* (Amsterodami: I. Hondius)

Mercator, Gerardus et I. Hondius, 1638, *Atlas Novus sive descriptio geographica totius orbis terrarum tabulis aeniis luculentissimis & accuratissimis exornata tribus tomis distinctus* (Amstelodami apud Ioannem Ianssonium & Henricum Hondium)

Mercator, Gerhard, 1891, Drei Karten von Gerhard Mercator: Europa, Britische Inseln, Weltkarte: facsimile-Lichtdruck nach den Originalen der Stadtbibliothek zu Breslau, hergestellt von der Reichsdruckerei/hrsg. Von der Gesellschaft für Erdkunde zu Berlin (Berlin: W. H. Kühl)

Merrill, S., 1881, *East of the Jordan: A Record of Travel and Observation in the Countries of Moab, Gilead and Bashan* (London: R. Bentley)

Meurer, Peter H. (1988), 'Die Palästina-Karte Mark Jordans (vermutlich Wittenberg) um 1546', *Speculum Orbis* 1 (1987), 9–12

Michaud, M., 1844, 'de la Roque, J. J'., in *Biographie universelle et moderne* (nouvelle edition, Paris: Desplaces; Leipzig: Brockhaus), vol. 36

Miller, J. M., 2000, 'Burckhardt-Robinson features in nineteenth-century maps of the Kerak plateau', in L. A. Stager, J. A. Greene and M. D. Coogan (eds.), *The Archaeology of Jordan and Beyond; Essays in Honor of James A. Sauer* (Winona Lake, Indiana: Eisenbrauns), 351–66

Miller, Konrad, 1895–98, *Mappaemundi: die ältesten Weltkarten*, I–VI (Stuttgart: J. Roth)

Miller, Konrad, 1916, *Itineraria Romana: Romische Reisewege an der Hand des Tabula Peutingeriana* (Stuttgart: Strecker und Schröder)

Miller, Konrad, 1962, *Das Peutingersche Tafel* (Stuttgart: Brockhaus)

Miller, Konrad, 1981, *Weltkarte des Arabers Idrisi vom Jahre 1154* (Stuttgart: Brockhaus/Antiquarium)

Mitchell, R. J., 1964, *The Spring Voyage: The Jerusalem Pilgrimage in 1458* (London: J. Murray)

Mittmann, S., and Schmitt, G., 2001, *Tübinger Bibelatlas: auf der Grundlage des Tübinger Atlas des Vorderen Orients (TAVO)* (Stuttgart: Deutsche Bibelgesellschaft; *TAVO Karten*: Wiesbaden: Dr Ludwig Reichert Verlag)

More, J., [1683], 1929, 'Wall Map of Canaan, review of' in *Geographic Journal* (London: Royal Geographic Society) 73

Moreland, C., and D. Bannister, 1989, *Antique Maps* (3rd ed.; London & New York: Phaidon)

Moscrop, J. J., 2000, *Measuring Jerusalem: The Palestine Exploration Fund and British Interests in the Holy Land* (London & New York: Leicester University Press)

Münster, Sebastian, 1550, *Cosmographiae universalis libri VI* (Basileae: apud Henrichum Petri)

Münster, Sebastian, 1552, *Geographiae Claudii Ptolemaei Alexandrini. . .Libri VIII* (4th ed., Basel: apud H. Petrum)

Musil, A., 1906–1908, *Arabia Petraea* (3 vols, Kaiserliche Akademie des Wissenschaften, Wien: A. Hölder)

Musil, A., 1907, *Kusejr 'Amra* (2 vols, Kaiserliche Akademie des Wissenschaften, Wien: A. Hölder)

Nebenzahl, K., 1986, *Maps of the Holy Land: Images of Terra Sancta through Two Millennia* (New York: Abbeville)

Neumann, G.-A., 1885, *Anonymi Lubicensis Prologus Arminensis in Mappam Terrae Sanctae Lubicae ante annum MCDLXXX impressus* (Genevae: Typis J.-G. Frick)(facsimile edition)

Newman, H. E. M., 1949, 'The Ordnance Survey in the Near East', *Royal Engineers Journal*, March-June, 64–77

Nordenskiold, A. E., 1889, *Facsimile Atlas to the early History of Cartography with Reproductions of the most important Maps printed in the XV and XVI Centuries* (Stockholm: A. A. Norstedt; reprinted New York: Dover Books, 1973)

North, R., 1979, *A History of Biblical Map Making* (Beihefte zum Tűbinger Atlas des vorderen Orients, Reihe B, 32. Wiesbaden: Reichart)

Noth, M., 1964, *The Old Testament World* (Eng. trans. of *Die Welt des Alten Testaments*, 4th ed., 1964, by V. I. Gruhn)(London: Black)

O'Loughlin, T., 2004, 'Two sixteenth-century biblical maps in The Founders' Library, University of Wales, Lampeter', *Trivium* 35, 121–68

O'Loughlin, T., 2005, 'Map and text: A mid ninth-century map for the Book of Joshua', *IM* 57.1, 7–22.

Olivier, J. P. J., 1995, 'Ulrich Jasper Seetzen: a pioneer in need of recognition', in M. Weippert and S. Timms (eds), *Meilenstein: Festgabe fur Herbert Donner zum 16 Februar 1995* (Wiesbaden: Harrassowicz), 164–71

Ortelius, A., 1579, *Nomenclator Ptolemaicus*, in id., *Theatrum Orbis Terrarum* (Antverpiae: Plantin)

Ortelius, A., 1570, 1579, 1589, *Theatrum Orbis Terrarum* (Antverpiae: Plantin)

Ortelius, a., 1587, *Thesaurus geographicus* (Antverpiae: Plantin)

Ortelius, A., 1600, *Itinerarium Antonini Augusti et Burdigalense AD CLV* (Coloniae Agrippinae: Birckmann)

Ortelius, A., 1624, *Theatri Orbis Terrarum Parergon, sive veteris geographiae tabulae commentariis et historicis illustratae, editio novissima* (Antverpiae: Plantin)

Pagninus, Santes, 1542, *Biblia Sacra in Santis Pagnini tralatione* (Lugdunum: apud Hugonem a Porta)

Palmer, E. H., 1872, *The Desert of the Exodus: Journeys on Foot in the Wilderness of the Forty Years' Wandering* (2 vols, Cambridge: Deighton Bell)

Paulus, H. E. G., 1784, *Atlas Antiquus Danvillianus*, Norimbergae in officina C. Weigelio -Schneideriana

Peile, J., 1900, *Christ's College* (University of Cambridge College Histories, London: F. E. Robinson)

Piccirillo, M., 1992, *The Mosaics of Jordan* (ed. P. M. Bikai and T. A. Dailey) (ACOR Research Publications 1. Amman: ACOR.)

Piccirillo, M., and E. Alliata (eds), *The Madeba Map Centenary 1897–1997: Travelling through the Byzantine Ummayad Period* (SBF Collectio Maior 40. Jerusalem: Studium Biblicum Franciscanum)

Pococke, R., 1743, 1745, *A Description of the East and some Other Countries. 1. Observations on Egypt; 2. Observations on Palestine or the Holy Land, Syria, Mesopotamia, Cyprus and Candia* (London: printed for the author)

Porter, J. L., 1855, *Five Years in Damascus: including an Account of the History, Topography and Antiquities of that City, with Travels and Researches in Palmyra, Lebanon and the Hauran* (2 vols, London: J. Murray)

Porter, J. L., 1856, 'Memoirs on the map of Damascus, Hauran, and the Lebanon mountains constructed from personal survey', *JRGS* 26, 43–54

Porter, J. L., 1858, and 2nd ed., 1875, *Handbook for Travellers in Syria and Palestine* (London: J. Murray)

Prestcott, H. F. M., 1964, *Jerusalem Journey: Pilgrimage to the Holy Land in the Fifteenth Century* (London: Eyre and Spottiswoode)

Pringle, D., 1993 (vol. 1), 1998 (vol. 2), *The Churches of the Crusader Kingdom of Jerusalem: A Corpus.* (Cambridge University Press)

Prologus Arminensis in Mappam Terrae Sanctae, 1478 (Lubeck: Lucas Brandis)

Prontero, F., 2003, *Tabula Peutingeriana* (Le antiche vie del mondo)(Florence: L. S. Olschki)

Ptolemaeus, Claudius, 1490, *Geographiae Libri VIII* (Rome: Petrus de Turre)

Ptolemaeus Cl., 1618, *Geographiae Libri VIII Graeci et Latini* (ed. P. Bertius Beverus)(Amsterdam: I. Hondius)

Ptolemaeus, Claudius, 1901, *Geographia* I.2, ed. C. Muller (Paris: Firmin-Didot)

Reed, N., 1978, 'Pattern and purpose in the Antonine Itinerary', *AJP* 99, 228–51

Reiske, J., 1711, *Philippi Cluverii Introductio in universam geographiam locupleta additamentis et annotationibus John Frid. Hekelii et J. Reiskii* (London: typis M. Jenour impensis Ioannis Nicholsonii)

Rekers, B., 1972, *Benito Arias Montano (1527–1598)* (London: Warburg Inst.; Leiden: Brill)

Reland, Hadrianus (Adrian), 1714, *Palaestina ex monumentis veteribus illustrata et chartis geographicis accuratioribus ornata* (Traj. Batav.: G. Broedelet)

Review of John More's 'Wall Map of Canaan', *Geographical Journal* (London Royal Geographical Society) 73

Riley-Smith, J. (ed.), 1991, *The Atlas of the Crusades* (London: Times Books)

Ritter, K., 1848–55, *Vergleichende Erdkunde von der Sinai-Halbinsel, von Palästina u. Syrien,* I–IV (Berlin: G. Reimer)

Ritter, K. 1866, *The Comparative Geography of Palestine and the Sinaitic Peninsula*, transl. by W. L. Gage, vols. I–IV (Edinburgh: T. & T. Clark)

Robinson, E., and Smith, Eli, 1841, *Biblical Researches in Palestine, Mount Sinai and Arabia Petraea: A Journal of Travels in the Year 1838* (London: J. Murray)

Robinson, E., and Smith, Eli, 1856, *Later Biblical Researches in Palestine and the adjacent Regions in the Year 1852* (London: J. Murray)

Robinson, E., and Smith, E., 1867 , *Biblical Researches in Palestine and the adjacent Regions: A Journal of Travels in the Years 1838 and 1852* (3rd ed., London: J. Murray)

Röhricht, R., 1890, *Bibliotheca Geographica Palaestinae* (Berlin: H. Reuther; reprinted London 1963, 1989)

Röhricht, R., 1891, 'Karten und Pläne zur Palästinakunde aus dem 7 bis 16 Jahrhundert (Tafel 1)', *ZDPV* 14, 8–11

Röhricht, R., 1895, 'Karten und Pläne zur Palästinakunde aus dem 7–16 Jahrhundert (Tafeln 5–7)', *ZDPV* 18, 173–82

Röhricht, R., 1898, 'Marino Sanudo sen. als Kartograph Palästinas', *ZDPV* 21, 84–126

Röhricht, R., 1901, 'Die Palästinakarte Bernhard von Breitenbach's', *ZDPV* 24, 128–35 (Tafeln 1–3)

Röhricht, R., 1904, 'Die Palästinakarte des William Wey', *ZDPV* 27, 188–93 (Tafel VIII)

de la Roque, J., 1716, *Voyage de la 'Arabie heureuse* (Amsterdam: Steenhouwer & Utwerf)

de la Roque, J., 1718, *Voyage dans la Palestine* (Amsterdam: Steenhouwer & Utwerf)

de la Roque, J., 1722, *Voyage de Syrie et de Mont Libanon* (Paris: chez André Cailleau)

Rudimentum novitiorum sive chronicarum historiarum epitome, 1475 (Lubeck: Lucas Brandis)

Ruge, W., 1904–12, *Aelteres Kartographisches Material in deutschen Bibliotheken* (Erster, zweiter u.s.w. Bericht. Nachr. v.d. Königl. Gesellsch. d. Wiss. zu Göttingen)

Salmon, F. J., 1937, 'Palestine of the Crusades (a map of the country on scale 1: 350,000, with historical introduction and gazetteer)' (Jaffa: Survey of Palestine)

Sandys, George, 1615, 2nd ed., 1621, and later editions, *A Relation of a Journey begun An. Dom. 1610: Foure Books. Containing a Description of the Turkish Empire, of Egypt, of the Holy Land...* (London: W. Barrett)

Sanson, N., 1696, *Atlas Nouveau contenant toutes les parties du monde* (Paris: H. Jaillot)

Sanson, N., ed. J. Clericus, 1705, *Atlas antiquus sacer et profanus* (Amstelodami apud Petrum Mortier)

Sanson, N., 1711, *Geographia Sacra ex veteri et novo Testamento desumta* (Amstelaedami apud R. & G. Wetstanios)

Sanson, N., 1717, *Kanaan en d'omlegende Landen* (Leeuwarden: F. Halma)

Sanuto, Marino, 1321, *Liber Secretorum Fidelium Crucis super Terrae Sanctae Recuperatione et Conservatione.* V. sub Bongarsius, J, and Stewart, A.

Schumacher, G., 1886, *Across the Jordan* (London: R. Bentley)

Schumacher, G., 1888, *The Jaulan, surveyed for the German Society for the Exploration of the Holy Land* (translated by permission from the Transactions of the German Society for the Exploration of the Holy Land; London: R. Bentley)

Schumacher, G., 1890, *Northern 'Ajlun 'Within the Decapolis'* (London: PEF)

Schwarz, Th., 1888, *Uber den Verfasser und die Quellen des Rudimentum Novitiorum* (Rostock: Universitäts-Buchdruckerei von Adlers Erben)

Seeck, O. (ed.), 1876, *Notitia Dignitatum* (Berlin: Weidemann)

Seetzen, U. J., 1810, *A Brief Account of the Countries adjoining the Lakes of Tiberias, the Jordan and the Dead Sea* (Bath: published for the Palestine Association of London)

Seetzen, U. J., 1810, 'Uber die bei diesem Heft befordliche Karte von Palästina', and Charte von Palaestina reducirt aus dem von dem Herrn D. Seetzen an Ort und Stelle entworfenen Handzeichnungen, Gotha 1810', *Monatliche Correspondenz zur Beforderung der Erde- und Himmels-Kunde* 22, 542–51, and facing 722

Shalev, Zur, 2003, 'Sacred geography, antiquarianism and visual erudition: Benito Arias Montano and the maps in the Antwerp Polyglot Bible', *IM* 55, 56–80

Shaw, T., 1738, *Travels, or Geographical,Physical and Miscellaneous Observations relating to several Parts of Barbary and Levant* (Oxford)

Skelton, R. A., 1966, introduction to *Francesco Berlinghieri Geographia Florence 1482* (Atlas in Facsimile, Third Series, vol. IV. Cleveland and New York: World Publishing Company for Amsterdam: *Theatrum Orbis Terrarum*)

Smith, G. A., 1904, review of R. E. Brünnow and A. von Domaszewski, *Die Provincia Arabia, PEFQS* 36, 397–400

Smith, G. A., 1915, 2nd ed. 1936, *Atlas of the Historical Geography of the Holy Land* (London: Hodder and Stoughton)

Speed, John, 1611, *The Theatre of the Empire of Great Britaine, presenting an exact geography of the kingdoms of England, Scotland and Ireland and the isle adioyning* (London)

Speed, J., 1627 and later editions, id., with *A Prospect of the Most Famous Parts of the World* (London: printed for T. Bassett & R. Chiswell); new edition 1676, enlarged by E. Phillips

Speed, John, 1611–1640, many editions, *The Genealogies recorded in the Sacred Scriptures according to every Family and Tribe with the line of our Saviour Christ observed from Adam to the Blessed Virgin Mary* (London)

Stephanus Beluzius, 1761, *Miscellanea, Tom.I* (Lucca: apud V. Junctinium)

Stevens, Henry N., 1908 (2nd ed.), *Ptolemy's Geography: A brief Account of all the printed Editions down to 1730* (London: Stevens, Son & Stiles); 1972 reprint by Theatrum Orbis Terrarum Ltd, Meridian Publishing Co, N. Israel publisher, for their friends at the occasion of the New Year (Amsterdam)

Stewart, A. (trans., from Tobler's text), [1894], 1897, *John Poloner: Description of the Holy Land* (London: PPTS, vol. 6)

Stewart, A. (trans.), [1895], 1897, *Ludolph von Suchem: Description of the Holy Land and of the way thither written in the year A.D. 1350* (London: PPTS, vol. 12)

Stewart, A. (trans.), [1896], 1897, *Burchard of Mount Sion* (London: PPTS, vol. 12)

Stewart, A. (trans.), [1896], 1897, *Marino Sanuto: Liber secretorum fidelium crucis*, Pt xiv of Bk III, with geographical notes by C. R. Conder (London: PPTS, vol. 12)

Stewart, A. (trans.), [1890], 1897, *Description of the Holy Land. John of Wurzburg (A.D. 1160–1170)*(London: PPTS, vol. 5, with notes by C. W. Williams)

Stewart, A., (trans), [1891], 1897, *Theoderich's Description of the Holy Places (c. 1172 A.D.)*(London: PPTS, vol. 5)

Stone, J. R., 1996, 'The medieval Mappaemundi: toward an archaeology of sacred cartography', *Religion* 23, 197–216

Talbert, R. J. A. (ed), 2000, *Barrington Atlas of the Greek and Roman World* (Princeton, NJ: Princeton U. P.)

Thomsen, P., 1906, 'Untersuchungen zur älteren Palästinaliteratur', *ZDPV* 20, 101–32

Thomsen, P., 1907, *Loca Sancta: Verzeichnis der im 1. bis 6 Jahrhundert n. Chr. erwähnten Ortschaften Palästinas mit besonderer Berücksichtigung der Lokalisierung der biblischen Stätten, 1 Band mit einer Karte* (Halle a. S.: Haupt)

Tishby, A., ed., 2001, *Holy Land in Maps* (Jerusalem: The Israel Museum)

Tooley, R. V., 1978, 'Maps of Palestine in the Atlas of Ortelius', *The Map Collector*, June, 28–31

Tristram, H. B., 1873, *The Land of Moab* (London: J. Murray)

Tsafrir, Y., 1986, 'The maps used by Theodosius: on the pilgrim maps of the Holy Land and Jerusalem in the sixth century C. E.', *Dunbarton Oaks Papers* 40, 129–45

Tudeer, L. O. Th., 1917, 'On the origins of the maps attached to Ptolemy's geography', JHS 37, 62–76

Undique ad Terram Sanctam: Cartographic Exhibition from the Eran Laor Collection, 1976 (Jerusalem: Jewish National and University Library)

Van der Steen, E. J., 2006, 'Nineteenth-century travellers in the Wadi Arabah', P. Bienkowski and K. Galor (eds), *Crossing the Rift: Resources, Routes, Settlement Patters and Interaction in the Wadi Arabah* (Oxford: Oxbow Books)

Vaughan, R., 1958, *Matthew Paris* (Cambridge Studies in Medieval Life and Thought, n.s.6; Cambridge: Cambridge University Press)

van de Velde, C. W. M., 1854, *Narrative of a Journey through Syria and Palestine in 1851 and 1852* (2 vols, Edinburgh and London: Blackwood)

van de Velde, C. W. M., 1858, 'Map of the Holy Land' (Gotha: Justus Perthes)

van de Velde, C. W. M., *Memoir to accompany the Map of the Holy Land constructed by C. W. M. van de Velde* (Gotha: Justus Perthes)

van 't Hoff, R. (ed.), 1961, *Gerard Mercator's* Map of the World *(1569) in the Form of an Atlas in the Maritime Museum 'Prins Hendrik' at Rotterdam* (Rotterdam/'s-Gravenhage: Maritiem Museum), *IM*, Supplement 2

Visscher, C. J., 1642, *Geographische Beschryvinge van T' Beloofde Landt Canaan* (Amstelodami)

de Vogüé, C. J. M., 1860, *Les églises de la Terre Sainte* (Paris: Didron)

Volney, C. F., 1787, *Voyage en Syrie, et en Egypte pendant les annees 1783, 1784, 1785* (2 vols, Paris: Desenne; Eng. ed., 2 vols, London: G. G. J. & J. Robinson, 1787)

Wajntraub, E. and G., 2005, 'Jacob Ziegler: an enlightened man from a dark period', *Journal of the International Map Collectors' Society* 100, 49–54

Waldeseemüller, M., 1513, *Ptolemaei Geographia* (Strassburg: Schott)

Walton, B., 1657, *Biblia Sacra Polyglotta* (London: T. Roycroft)

Warren, C., 1880, *Plans, Elevations, & Sections shewing the results of the Excavations at Jerusalem 1867–70, executed for the Committee of the PEF by Captain Charles Warren RE* (London: Committee of the PEF) (Plate XLVIII, 'Reconnaissance of the Jordan Valley by Lieut. Charles Warren, RE, 1867')

Warren, C., 1884, *Survey of Western Palestine* (London: Committee of the PEF)

Warren, C., 1869–70, 'Expedition to East of Jordan, July and August 1867', *PEFQS*, 284–311

Warren, C., 1869–70, 'Positions of places east of Jordan', *PEFQS*, 381–88

Watelet, M. (ed.), 1994, *Gerard Mercator cosmographe* (Antwerp: Fonds Mercator)

Watelet, M., 1998, *The Mercator Atlas of Europe: Facsimile of the Maps by Gerardus Mercator contained in the Atlas of Europe, circa 1570–1572* (Pleasant Hill OR: Walking Tree Press)

Weippert, Helga, 1995, 'Unterwege nach Afrika: Ulrich Jasper Seetzen (1767–1811)', in M. Weippert and S. Timms (eds), *Meilenstein: Festgabe fur Herbert Donner zum 16 Februar 1995* (Wiesbaden: Harrassowicz)

Wey, William, 1857, *Itineraries to Jerusalem A.D. 1458 and 1462* (London: Roxburghe Club)[introduction by G. Williams]

Wey, William, 1867, *Map of the Holy Land illustrating his itineraries in 1458 and 1462* (London: Roxburghe Club)

Whitfield, P., 1994, *The Image of the World* (London: British Library)

Wilkinson, J., 1977, *Jerusalem Pilgrims before the Crusades* (Warminster: Aris and Phillips)

Wilkinson, J, J. Hill & W. F. Ryan, 1988, *Jerusalem Pilgrimage 1099–1195* (London: Hakluyt Society)

William of Tyre. See under Huygens, R. B. C.

Wilson, C. W., 1873, 'Recent surveys in Sinai and Palestine', *JRGS* 43, 213

Wilson, C. W., R.E., and H.S. Palmer, R.E., 1869–72, *Ordnance Survey of the Peninsula of Sinai, I–V* (Southampton: Ordnance Survey)

Winter, H., 1952, 'Petrus Roselli', *IM* 9, 1–11

Wissenburg, Wolfgang, 1535, *Terrae Sanctae altera descriptio, iuxta ordinem alphabeti, quae ad Scripturam proxime directa est, utilissima etiam plebeio lectori* (Argentatori apud Vuendelium Rihelium)

Wissenburg, Wolfgang, 1538, *Descriptio Palestinae nova* (Basel)

Wit, Frederick de, [n.d.; late 17th century], *Atlas*, and *Atlas Maior* (Amsterdam: de Wit)

Ziegler, Jacob, 1532, *Quae intus continentur Syria. . ., Palestina. . ., Arabia etc.* (Argentatori [Strassburg]: apud P. Opilionem)

Ziegler, Jacob, 1536, *Terrae Sanctae, quam Palestinam nominant. . . . .descriptio* (Argentatori: apud Vuendelinum Rihelium)

Zimmermann, Kl., 2003, 'Cartography', *New Pauly* (ed. C. F. Salazar; Eng. ed. H. Cancik and H. Schneider) (Brill: Leiden), 1138–44

INDEXES

I. AUTHORS

II. INDEX OF PLACE NAMES

NB. This list includes names given in the text, but not names from the tables or the maps. Spellings vary considerably according to author and language, and the most important, but not all, variant spellings are here included. In the main text, in the discussion of any particular author, I have used the form of the names given by that author.

For Product Safety Concerns and Information please contact our EU representative GPSR@taylorandfrancis.com Taylor & Francis Verlag GmbH, Kaufingerstraße 24, 80331 München, Germany

Batch number: 10722116

Printed by Printforce, the Netherlands